Chet Raymo is a professor of physics and astronomy
in North Easton, Massachusetts. He has lectured to Scouts,
clubs, church groups, and school children
on the subject of stars.

PHalarope Books are designed specifically for the amateur
naturalist. These volumes represent excellence in natural history
publishing. Each book in the PHalarope series is based on a nature
course or program at the college or adult education level or is
sponsored by a museum or nature center. Each PHalarope Book
reflects the author's teaching ability as well as writing ability.

BOOKS IN THE SERIES:

The Curious Naturalist
John Mitchell and the
Massachusetts Audubon Society

The Art of Painting Animals:
A Beginning Artist's Guide to the
Portrayal of Domestic Animals,
Wildlife, and Birds
Frederic Sweney

The Amateur Naturalist's Handbook
Vinson Brown

Nature Drawing: A Tool for Learning
Clare Walker Leslie

Outdoor Education: A Manual for
Teaching in Nature's Classroom
Michael Link, Director,
Northwoods Audubon Center,
  Minnesota

The Wildlife Observer's Guidebook
Charles E. Roth,
Massachusetts Audubon Society

Nature with Children of All Ages:
Activities and Adventures for Exploring,
Learning, and Enjoying the World
Around Us
Edith A. Sisson,
Massachusetts Audubon Society

Nature Photography: A Guide to
Better Outdoor Pictures
Stan Osolinski

A Complete Manual of Amateur
Astronomy: Tools and Techniques
for Astronomical Observations
P. Clay Sherrod with Thomas L. Koed

365 Starry Nights: An Introduction to
Astronomy for Every Night of the Year
Chet Raymo

STARRY NIGHTS

AN INTRODUCTION TO ASTRONOMY
FOR EVERY NIGHT OF THE YEAR

TEXT & ILLUSTRATIONS BY CHET RAYMO

SPECTRUM BOOK

Prentice-Hall, Inc., Englewood Cliffs, New Jersey 07632

Library of Congress Cataloging in Publication Data

RAYMO, CHET.
   365 starry nights.

   (PHalarope books)
   "A Spectrum Book."
   1. Astronomy—Observers' manuals.   1. Title.
QB64.R38   523   82-7511
ISBN 0-13-920520-9   AACR2
ISBN 0-13-920512-8 (pbk.)

© 1982 by Prentice-Hall, Inc., Englewood Cliffs, N.J. 07632

A Spectrum Book

Printed in the United States of America

10 9 8 7 6 5 4 3 2 1

ISBN 0-13-920520-9

ISBN 0-13-920512-8 (PBK.)

This Spectrum Book can be made available to businesses and organizations at a special discount when ordered in large quantities. For further information contact Prentice-Hall, Inc., General Publishing Division, Special Sales, Englewood Cliffs, N.J. 07632.

Editorial/production supervision and
   interior design by Maria Carella
Cover design by Jeannette Jacobs
Manufacturing buyer: Cathie Lenard

Prentice-Hall International, Inc., *London*
Prentice-Hall of Australia Pty. Limited, *Sydney*
Prentice-Hall of Canada Inc., *Toronto*
Prentice-Hall of India Private Limited, *New Delhi*
Prentice-Hall of Japan, Inc., *Tokyo*
Prentice-Hall of Southeast Asia Pte. Ltd., *Singapore*
Whitehall Books Limited, *Wellington, New Zealand*

Contents

For Tom . . .

a welcome companion
on starry nights
and a resourceful collaborator
in the making of the book.

Introduction

Let it be said at once that although I have been trained as a scientist and have taught courses in descriptive astronomy, my interest in the sky is primarily esthetic rather than scientific. If I were to be exiled on a desert island and allowed to take the traditional handful of books, they would not be works of science but of poetry and natural history. This prejudice is revealed in the title of this book. The title refers not to *stars* but to *starry nights*. Even as I write the words I am reminded of Van Gogh's magical painting *The Starry Night*. Van Gogh's vision of the sky over Saint-Rémy, alive with swirling dragon nebula, moonlike stars, and sunlike moon, sweeps us deep into a starry night that can only be dreamed. The night sky pos-

sesses an unparalleled power to excite the human imagination. Intimate, yet infinite. Dark, yet full of light. Near, yet unreachably far. No part of our world displays such immediately accessible patterns of order, and no part of our world remains so deeply mysterious. According to the anthropologists, our ability to count, to factor time, to measure space, to invent myths, and to do rational science are closely bound up with our ancestral experience of the sky.

This book is designed to be a kind of companion to the night. It is full of science, but only because (as the old catechisms used to say) knowledge is a prerequisite for love. Knowing the night sky is a different thing from knowing, say, the mechanism of a clock or a computer. The clock or the computer is finite, to know it is to exhaust its potential for exciting wonder. The night sky is more like a human being, inexhaustibly complex and finally beyond reach. Knowledge only whets our interest and increases our wonder.

The selection of things to be included in this book is entirely personal. For example, the reader of an early draft of this book commented on the number of times I mention tiny red dwarf stars which just happen to be near neighbors of the sun. Who cares, after all, about a star named Ross 154, which is not much larger than Jupiter and is invisible to the unaided eye? These little red stars receive scant notice in star books, yet they are the most numerous denizens of the Galaxy. I include them to redress the balance.

The book is meant for the naked-eye observer but often treats of things that can be observed only with the largest observatory telescopes. "The power of the visible," says Marianne Moore in one of her poems, "is the invisible." No more of the universe is visible to our unaided eyes than to the eyes of our Neanderthal ancestors. But science, the product of our imagination, has immensely extended the range of our imagination. Our inward eye can range beyond the dome of visible stars to the unseen realm of the nebulae and galaxies.

I have confined myself to that part of the celestial sphere which can be observed from mid-northern latitudes. Someday I would like to spend 365 starry nights under southern skies, but for the time being I think it best to stay with that part of the sky I know best. Most of the maps and drawings in the book have been prepared for a hypothetical observer at a latitude of about 40° north. Still, the maps and drawings should prove entirely satisfactory for any observer in the United States, southern Canada, Europe, Japan, or other mid-northern locality. The drawings of clusters, galaxies, and star fields are meant to be suggestive only; major stars have been carefully placed, but the reader should not take every sprinkling of tiny dots literally. Several fine star atlases are recommended in the bibliography ("Sources and Resources").

The five naked-eye planets are the "movable feast" of the starry night. Anyone who watches the sky will want to keep track of their comings and goings. They are not included in this book precisely because they *are* a variable treat, changing their aspect and location from year to year. Your newspaper will probably tell you where to look for them.

Astronomy is not always an exact science. The distance to a star or the age of the universe cannot be determined with the same degree of accuracy as one can weigh the reagents in a chemical reaction. As I read the first draft of the book, I was struck by the number of times I had used the words "about" and "approximately." I deleted many of them. Bear in mind that if the distance to a nebula (for example) is given as 800 light years, the actual distance could turn out to be 600 or 1000. Often the authorities upon whom I relied gave very different values for the properties of celestial objects. I have tried to relate the best and most up-to-date information. Pronunciations of star names is another matter of some diversity.

It should be possible to pick up this book and begin reading on almost any page. Terms that may not be familiar are printed in boldface type and you will find them defined in the glossary, together with helpful cross-references. If you make it through all 365 starry nights, you will have completed a kind of mini-course in descriptive astronomy.

The format and spirit of the book are, I trust, unique. But the information contained here is entirely derivative and has been compiled from many sources. In the section called "Sources and Resources" I describe and recommend several of the books that I think you might like best. I would especially mention two authors, Robert Burnham and Guy Ottewell. Their books are gold mines of information and wonderfully original. Every serious stargazer will want to own them.

ACKNOWLEDGMENTS

I would like to thank Mike Horne, who read the manuscript; Maureen Raymo, who acted as editor and typist; and Maurice Sheehy, who came up with a ream of typing paper at a time and in a place I would have thought impossible.

I would also like to thank Mary Kennan, Editor, and Maria Carella, Senior Designer, of Prentice-Hall, who offered the skillful and steady hands that brought this book into existence.

The following illustrations have been adapted from figures in *Burnham's Celestial Handbook* by Robert Burnham (Vols. I and II, 1978; Vol. III, 1979), with the kind permission of the publisher, Dover Publications, Inc.:

Proper motion of Procyon, February 17

Light curve for U Geminorum, February 27

Orbits for Zeta Cancri, March 17

Light curve for R Leonis, April 10

Coin showing Berenice, April 27

Light curves for R and T Corona, June 19–20

Light curves for Beta and RR Lyrae, August 8, 11

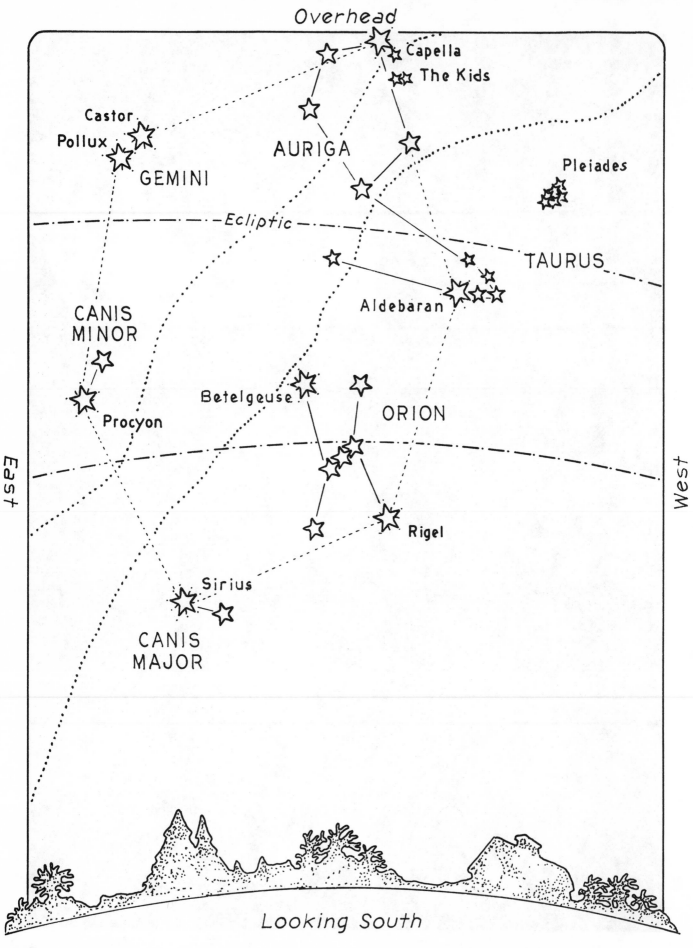

Overhead

Capella

The Kids

Castor
Pollux

GEMINI

AURIGA

Pleiades

Ecliptic

TAURUS

CANIS
MINOR

Aldebaran

Betelgeuse

ORION

Procyon

East

West

Rigel

Sirius

CANIS
MAJOR

Looking South

2

JANUARY 1

JANUARY 2

1st: The map at the left shows all of the stars we will study during January—and more. To find them in the sky, it is best to start with Orion, one of the most conspicuous constellations. The stars of Orion vividly suggest the mythological figure they are supposed to represent—a bold hunter armed with a club and sword and faced by a charging bull. To observe Orion, find a place where you have a clear view of the sky. Turn so that you are facing south. During the evening hours this month you will find Orion about halfway up from the horizon to the zenith. The **zenith** is the point in the sky directly above your head. The most striking feature of the constellation is the alignment of three equally bright stars in the hunter's belt. If you are looking in the right direction you can't miss it. As you stand facing giant Orion, the glittering yellow star almost directly over your head is Capella in the constellation Auriga the Charioteer. Capella, with Aldebaran in Taurus the Bull, Rigel in Orion,

Sirius and Procyon in the Big and Little Dogs, and the Gemini twins Castor and Pollux, make up the *Winter Hexagon*. Betelgeuse, the brilliant red star in the arm of Orion, is near the center of the Hexagon. In January and February we will look closely at each of these stars and constellations in turn.

2nd: Once you have found and learned the stars of Orion's belt, you will never again have trouble recognizing this constellation. Look also for the bright stars of the shoulders and feet and the fainter stars of the head and sword. Orion rises in the early evening at the beginning of December and dominates the sky all winter long. The hunter's stars, which include two of the brightest in the heavens, outshine those of any other constellation. Because Orion stands above the earth's equator, it is visible from every inhabited place on earth. All human cultures in every time and place have given special note in story and myth to this wonderful array of stars. Sailors of old feared the sight of Orion, for his appearance on the eastern horizon forecast stormy winter weather. But the hunter has also long been associated in myth with the forces of goodness and light, and as such we welcome his appearance as a promise of sparkling starry nights to come.

JANUARY 3

JANUARY 4

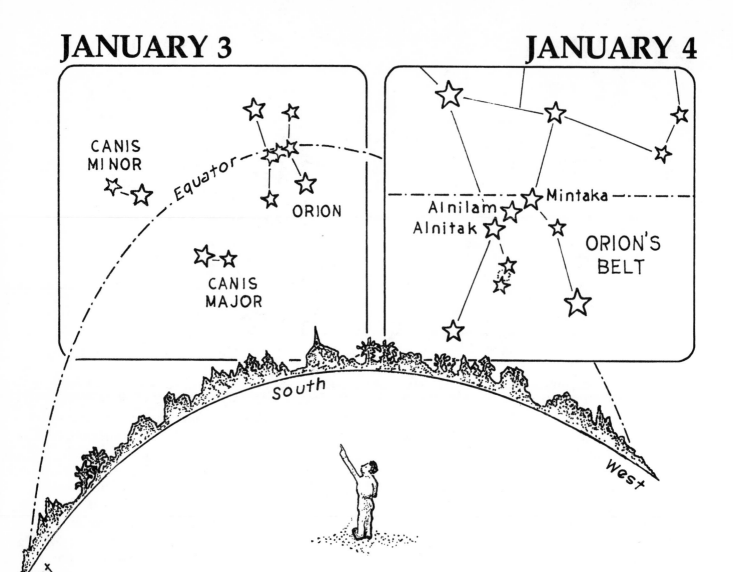

3rd: It is important to acquire early a sense of your own place under the stars. Orion's belt will help. Point to the place on your horizon that is due east of your observing location (a map of your town will help you find the approximate compass points—later you will use the stars). Now swing your arm up in an arc through the stars of Orion's belt, and on to the point on your horizon that is due west. You have traced out the sky's equator, or celestial equator. The **celestial equator** is the imaginary line among the stars that lies directly above the equator of the earth.

4th: It is convenient to imagine, as the ancients believed, that all of the stars lie on one great sky-sphere that surrounds and encloses the earth. We call this imaginary sphere the **celestial sphere.** In fact, as we know, the stars are distributed in space at different distances from earth, and there is no "sky-sphere." The three stars of Orion's belt, however, do lie at about the same distance from the earth and are therefore part of a true cluster. Like many of the stars, the stars of the belt have Arabic names. These names derive from the time of Europe's "Dark Ages" when the Arabs were the

keepers and developers of ancient Greek astronomy. When Europeans rediscovered astronomy during the late Middle Ages, it was often by way of Arabic translations of the Greek texts. Mintaka *(MIN-tack-a)* means "belt." Alnitak *(Al-NYE-tack)* also means "the belt." The meaning of Alnilam *(Al-NILE-am)* is less clear but is possibly "the belt of pearls," a beautiful name for this string of dazzling stars. As we go along, you may begin to wonder why so many star names begin with "Al-." The answer is a simple one. "Al-" is the Arabic prefix which means "the."

JANUARY 5

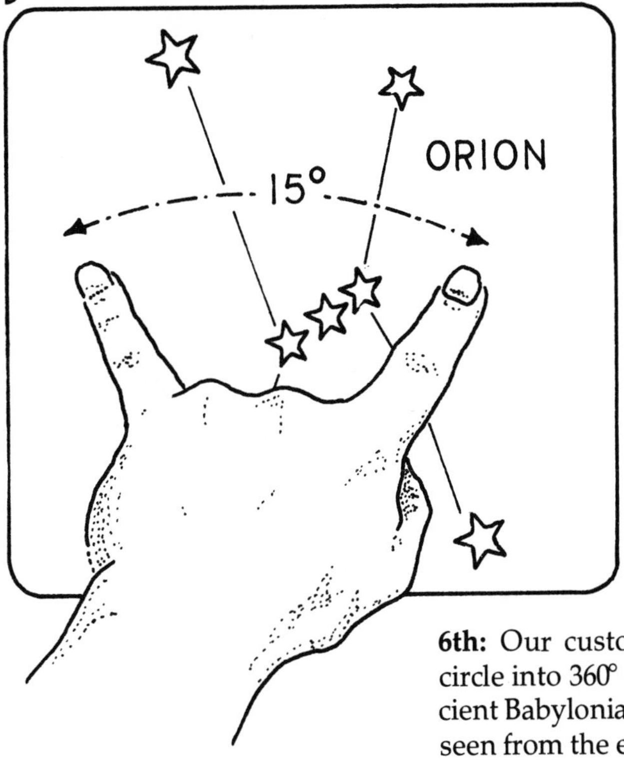

ORION

15°

JANUARY 6

1 degree

Betelgeuse

○ Moon
½°

5°

10°

5th: We require a convenient way to describe the positions of the stars on the **celestial sphere.** This is accomplished through the use of an *angular method of measurement,* with the vertex of the measured angle at the eye of the observer. The full circle of the sky, all the way around the earth, is 360 degrees (360°). The angle from horizon to horizon passing over your head is 180°, and the angle from the horizon to the zenith is 90°. If you stretch your arm out in front of you and sight along it with one eye, the angle between the tips of your spread fingers is about 15°. This is the width of the constellation Orion. It should take about six handspans to measure the distance from the horizon to the zenith. Try it and see how closely your hand and arm fit the rule.

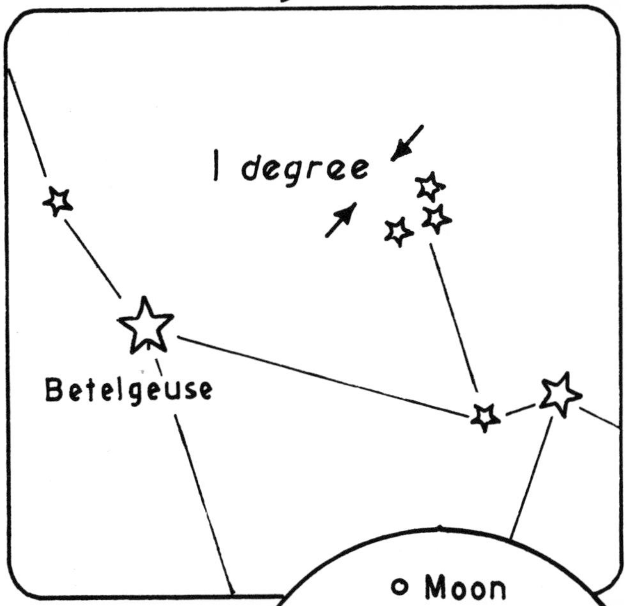

6th: Our custom of dividing a circle into 360° derives from ancient Babylonian astronomy. As seen from the earth, the sun appears to make a full circuit of the sky in a little over 365 days (see Jan. 8). A degree, then, as defined by the Babylonians, was about the distance the sun moved each day with respect to the background of stars. The three little stars of Orion's head occupy a circle of about 1° diameter. The angular size of the moon against the sky is ½°, or about half the width of your little finger held at arm's length. The moon would therefore fit nicely between the three stars of Orion's head. If you look for these stars on a clear night you might have the impression that the moon is much larger than the space they enclose. Hold out your little finger against these stars and then against the moon, and you will discover that the moon is smaller than you think. Some other useful guides for measuring angles in the sky are shown at the right.

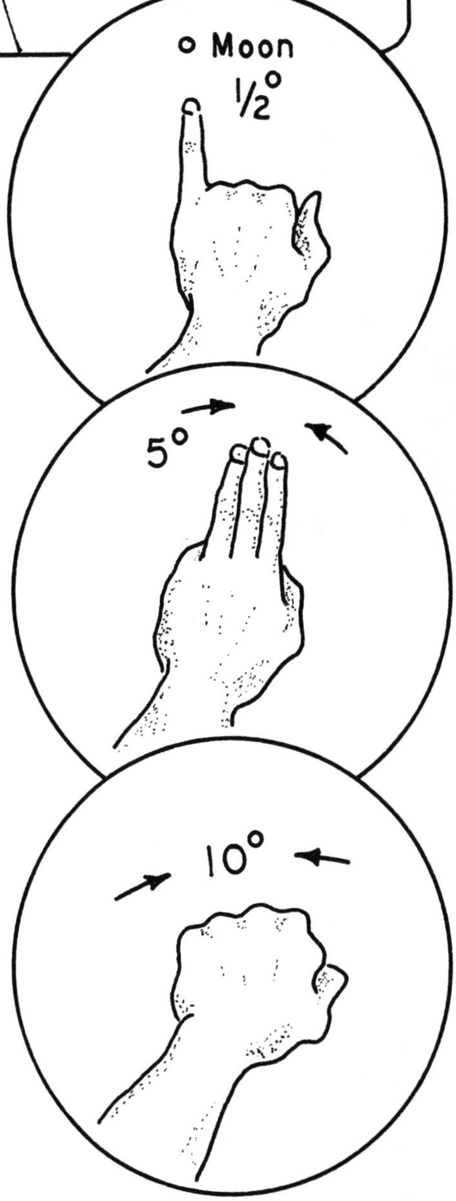

5

JANUARY 7

JANUARY 8

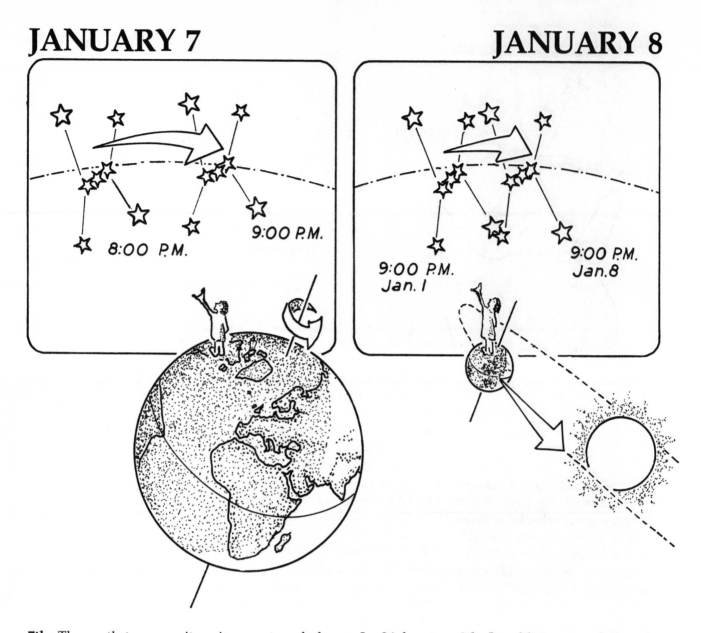

7th: The earth turns on its axis under the stars once every 24 hours, and carries us around as it goes. The stars remain fixed in the deeps of space. The earth turns west to east on its axis. As a result, the stars—with sun, moon, and planets—*seem* to move from east to west, making one full circuit around the earth each day. Like the sun and moon, the stars of Orion rise in the east and set in the west about 12 hours later. If you watch Orion throughout the evening, you will see him move one handspan (15°) toward the

west each hour. In 24 hours, Orion will set in the west, pass beneath the earth, and rise again from the east to regain his present position. The illusion that it is the stars, not us, which move is very powerful. The earth is near at hand and seems massive and stationary compared to the apparently tiny celestial objects. Only since the brilliant theoretical work of Nicholas Copernicus in the 16th century have we come to recognize that the "turning" of the stars is actually the turning of the earth.

8th: In addition to a daily spin on its axis, the earth makes a great annual journey around the sun. Since the stars we see at night are those on the side of the earth opposite the sun, the evening sky changes as our vantage point changes. As the earth carries the observer eastward around the sun, the stars seem to move night by night toward the west at a rate of about 1° per day. In 6 months' time, looking out into space from the other side of the sun, we shall find other stars in our starry night.

JANUARY 9

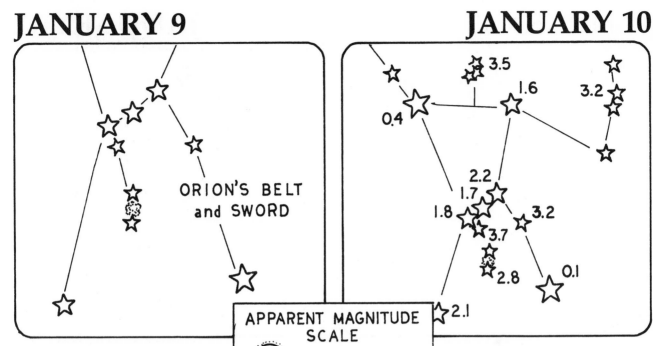

ORION'S BELT
and SWORD

JANUARY 10

3.5

1.6

3.2

0.4

2.2

1.7

1.8

3.2

3.7

2.8

0.1

2.1

APPARENT MAGNITUDE SCALE

Sirius, Canopus

-1

Rigel, Capella, Arcturus, Vega

0

Betelgeuse, Aldebaran

1

Alnilam, Alnitak, Mintaka, Nath

2

Orion's knee

3

Stars of Orion's head

4

Hundreds of stars

5

Thousands of stars

6

Greece

Mediterranean Sea

Alexandria

Egypt

9th: The stars do not appear equally bright in the sky. This is due to two things: (1) the stars are at different distances from the earth, and (2) the stars are not all of the same intrinsic brightness. The scale that is used to describe the brightness of stars as they appear to earth observers is called the scale of **apparent magnitude.** The scale was invented by the astronomer Hipparchus who lived and worked in the city of Alexandria 2100 years ago. The brightest stars in the sky, like Rigel and Betelgeuse in Orion, Hipparchus called stars of the *first magnitude*. The faintest stars he could see, he called *sixth-magnitude* stars. To other stars he assigned appropriate magnitudes between these limits. Thousands of years later we still use Hipparchus' scale of apparent brightness, although it has of course been made quantitatively more exact. Hipparchus was one of several great astronomers of the ancient world who was associated with the city of Alexandria.

10th: The exact modern magnitudes of the stars of Orion are shown above. If you live near city lights you will not see stars less bright than about the fourth magnitude, or no more than several hundred stars at any one time. If you live where the sky is very dark, on a clear night you might see several thousand stars down to the sixth magnitude. The stars of Orion's head are about the fourth magnitude and are a good test of the quality of the night. Of course, with binoculars or a telescope, you can see many more stars than could be seen by Hipparchus, even on dark Alexandrian nights unmarred by atmospheric pollution or electric lights. With the invention of the telescope it was necessary to extend the scale of apparent magnitude to encompass stars less bright than the sixth magnitude (see Oct. 25–28). At the other end of the scale, a few stars in the sky—Sirius, Canopus, Alpha Centauri, and Arcturus—have been assigned negative magnitudes on the modern scale.

7

JANUARY 11

JANUARY 12

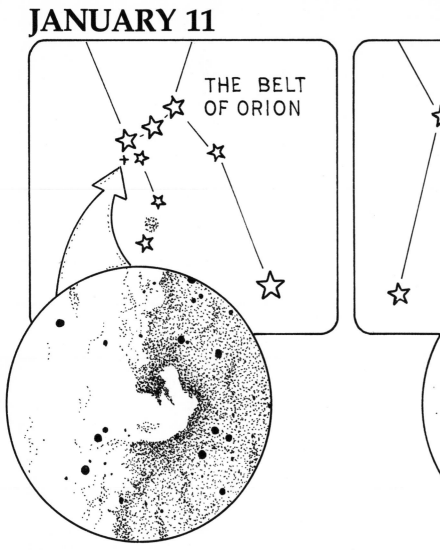

THE BELT OF ORION

THE SWORD OF ORION

11th: Just below the star Alnitak in Orion's belt is one of the most famous objects in the sky. The *Horsehead Nebula* is a dark cloud of dust and gas silhouetted against a brighter region of glowing interstellar gas heated to incandescence by the energy of the many stars embedded within it. The Horsehead takes its name from its shape. The size of this dark cloud is almost too great to imagine. A billion solar systems would fit neatly inside, and the Horsehead is just a wisp of a much larger cloud! You will not see the Horsehead with the naked eye. It is best seen with long-exposure telescopic photographs.

12th: The vast spaces between the stars are filled with dust and gas. It is in just such dense nebulae as the Horsehead that astronomers believe stars and possibly planets are born, condensing by gravity from the material of the cloud. If a knot of condensing gas—it is mostly hydrogen—is compressed to a density and temperature great enough for nuclear fusion to occur (see Jan. 14), a star is born. This is the process that is occurring even now in the *Great Orion Nebula*. What looks to the naked eye like three stars in Orion's sword are revealed by a small telescope to be a fascinating complex of stars and glowing gas clouds. Even to the unaided eye the central "star" of the sword may seem fuzzy. This "star" is the Great Nebula, a spectacular region of turbulent gas and dust heated by the radiation of hot young stars that have only recently (by star time!) condensed from the matter of the cloud. The diameter of the Great Nebula is more than 20,000 times that of the solar system, and there is enough hydrogen, helium, and other materials in the cloud to form at least 10,000 stars similar to our sun. Stars have lifetimes such as we do, although much longer ones. In the Great Orion Nebula we observe their births; shortly we will see how they die.

JANUARY 13

JANUARY 14

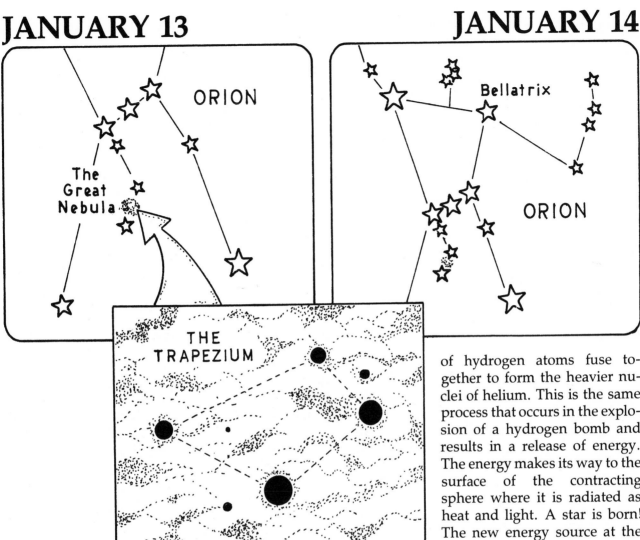

13th: Embedded in the very heart of the Great Orion Nebula, and visible with binoculars or a small telescope, is the beautiful multiple star system known as the *Trapezium,* four hot young stars in a tight trapezoid-shaped cluster. Actually, these four stars are only the brighter components of an expanding cluster containing hundreds of faint stars. The intense radiation from these high-temperature stars excites the gas of the surrounding nebula and makes it glow. To the eye the nebula glows with an eerie green light, but photographs show beautiful hues of pink, blue, and violet.

14th: From the expansion rate of the Trapezium cluster and from the color and brightness characteristics of the member stars (see Mar. 19–20), it has been estimated that the stars in the group may be less than half a million years old—making it one of the youngest associations of stars known. Indeed, some members of the group may even now be "turning on" their nuclear energy sources to become stars. As gravity pulls together a knot of gas and dust from the Great Nebula, the pressure and temperature at the core of the contracting cloud go up. When the temperature reaches about 10 million degrees Celsius, nuclei

of hydrogen atoms fuse together to form the heavier nuclei of helium. This is the same process that occurs in the explosion of a hydrogen bomb and results in a release of energy. The energy makes its way to the surface of the contracting sphere where it is radiated as heat and light. A star is born! The new energy source at the star's core stops the gravitational contraction of the star. The star can continue to burn steadily—an outward pressure sustained by nuclear fusion balanced against gravity—for as long as the hydrogen at the core holds out. The lifetime of a hot blue star such as Bellatrix might be as little as 10 million years. Bellatrix is larger than the sun and contains 10 times as much matter. Although it has more fuel than the sun, Bellatrix is a hotter star and "burns" its hydrogen at a faster rate. The slower-burning sun will probably have a lifetime of 10 or 15 billion years. Since our star is now about 5 billion years old, we have about at least 5 billion years to go!

9

JANUARY 15

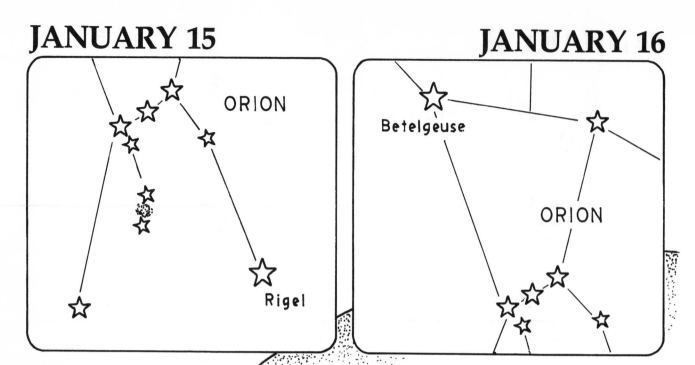

ORION

Rigel

JANUARY 16

Betelgeuse

ORION

15th: When a star such as Bellatrix has used up the hydrogen fuel at its core, gravity again gets the upper hand and the core of the star collapses. The sudden increase in pressure and temperature in the core releases new (but limited) sources of nuclear energy which heat the outer layers of the star and cause them to expand outward. As the star swells, the outer layers cool and change color from bluish, to white, to yellow, to red. The massive white star Rigel *(RYE-jell,* "foot") is possibly at this stage of its evolution and has begun the process of expansion. Rigel is 50 times bigger than the sun.

16th: If Rigel continues to expand, in tens of thousands of years it will become a **red giant** star like Betelgeuse *(BET-el-jews,* or *beetle juice* is close enough, "armpit of the giant"). Betelgeuse is one of the largest stars known. Its diameter is greater than the earth's orbit around the sun! It is one of the very few giant stars that has been seen as an actual disk, rather than a mere point of light. Using special techniques, astronomers have taken photographs that show features of its surface. The distinctive reddish color of Betelgeuse is readily apparent to the eye, particularly by contrast with white Rigel.

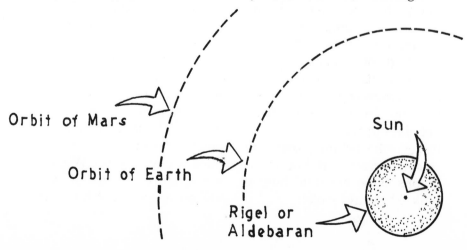

Betelgeuse

Orbit of Mars

Orbit of Earth

Rigel or
Aldebaran

Sun

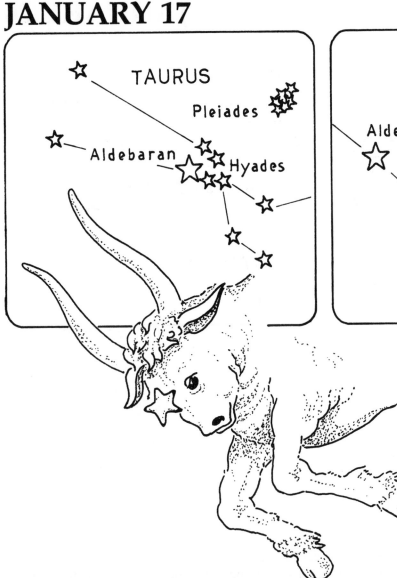

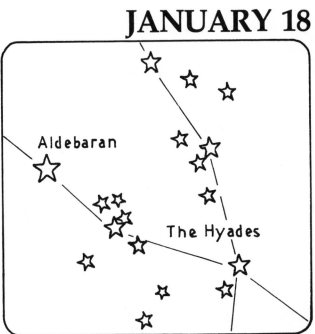

18th: Aldebaran *(al-DEB-a-ran)*, the bright red eye of the bull, is a **red giant** star with a diameter 36 times as great as the sun. Big—but not nearly so large as monstrous Betelgeuse! Aldebaran's Arabic name means "the follower." The star rises about one hour after the little group of stars known as the Pleiades (see Jan. 23), at almost the same place on the eastern horizon, and follows the cluster across the sky. The colors of stars are indicators of their temperatures, a relationship that is the same, say, as for a piece of iron heated from red hot to orange, to yellow, to white hot. Bluish or white stars like Rigel are very hot. Orange or yellow stars such as our sun are less hot. Ruddy stars like Aldebaran are the coolest of all. The intrinsic brightness of a star is determined by both its temperature and its size. Although cooler than the sun, Aldebaran would far outshine our star—if at the same distance from the earth—because of its much greater surface area.

17th: As he stands in the winter sky with his club raised, Orion faces a formidable opponent. Plunging toward him from the west is Taurus *(TOR-us)* the Bull, an awesome long-horned creature with threatening red eye and head lowered in a tumultuous charge. The bull is not so conspicuous a constellation as bright Orion, but the "vee" of stars that outlines the bull's face is easily recognized. Taurus is almost certainly one of the most ancient of the constellations. It is one of the familiar **signs of the zodiac,** the constellations that the sun moves through during its yearly journey across the sky (see June 25). Six thousand years ago, when astronomy and agriculture were developing together in the civilizations of the East and Near East, the sun was in Taurus on the first day of spring, a passage that marked the beginning of the cycle of planting, growth, and harvest. It is perhaps because of this that the bull figures so prominently in the myths and legends of the eastern Mediterranean worlds.

TAURUS

Ecliptic

Sun

Moon

Aldebaran

Ecliptic

TAURUS

19th: The ecliptic passes near Aldebaran. The **ecliptic** is the imaginary line on the celestial sphere that marks the sun's yearly journey through the stars. Of course, it is actually the earth that does the moving, not the sun. But as the earth travels around the sun, the sun appears to move against the background of the more distant stars. When the sun is in the sky, its great light, scattered through the earth's atmosphere, obliterates the fainter light of the stars. So we must imagine the background of stars in the sun's part of the sky. The sun is in the constellation Taurus in late May and early June.

20th: The full ecliptic passes through twelve constellations, collectively called the **zodiac.** The sun will always be found in one of these constellations. Because the moon and planets move more or less in the same flat plane as the earth's orbit (the **ecliptic plane**), we will always find them somewhere in the sky near the ecliptic. The "ecliptic" takes its name from the fact that an eclipse of the sun or moon can only occur when the moon is on the ecliptic and in a direct line with earth and sun. If you see a bright "star" in a zodiac constellation that is not on your star map, you are almost certainly looking at a planet. Watch the planet night by night and see it move with respect to the stars (the inner planets will move more rapidly than the outer ones). Sometimes the moon's monthly circuit around the earth takes it directly in front of a zodiac star, such as Aldebaran. When the moon blocks the light of the star we say that the star is *occulted* by the moon. An **occultation** of Aldebaran is an exciting event to watch with binoculars. Since the star is an almost perfect point of light and the moon has no atmosphere, the light of the star blinks out with surprising suddenness as the moon passes over it.

JANUARY 21

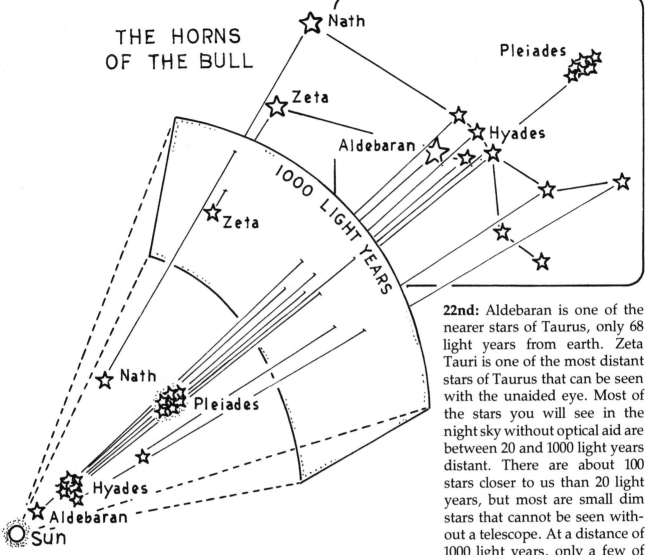

THE HORNS
OF THE BULL

21st: The stars we see as flat constellations on the celestial sphere are actually distributed in three-dimensional space. So stars which appear together in our sky may be very distant from each other in "real space." The two stars at the tips of the bull's horns, for example, are at very different distances from the earth. Nath ("butting horn") is about 300 light years away, and Zeta Tauri (*ZAY-tah TOR-ee*) is closer to 1000 light years distant, or three times further. The associations of stars we call constellations are therefore of mythological significance only, and have little physical meaning for the modern astronomer. They remain, however, convenient fictions for talking about the stars. A **light year** is the distance light travels in a year at a speed of 186,000 miles per second. The light by which we see Nath left that star 300 years ago! Zeta Tauri's light has been traveling toward us since the time of William the Conquerer. A light year is about 6,000,000,000,000 (6 trillion) miles.

22nd: Aldebaran is one of the nearer stars of Taurus, only 68 light years from earth. Zeta Tauri is one of the most distant stars of Taurus that can be seen with the unaided eye. Most of the stars you will see in the night sky without optical aid are between 20 and 1000 light years distant. There are about 100 stars closer to us than 20 light years, but most are small dim stars that cannot be seen without a telescope. At a distance of 1000 light years, only a few of the most luminous giant stars are still prominent. Many of the stars of Taurus are part of two great clusters. The Hyades (*HI-a-deez*), which include most of the stars of the bull's face, lie at a distance of about 130 light years. The Pleiades (*PLEE-a-deez*), a tiny twinkling cluster of stars in the body of the bull, are three times further from the earth. You will note from the drawing that an observer on a planet of Nath would not see the two clusters in the same part of the night sky.

JANUARY 23   # JANUARY 24

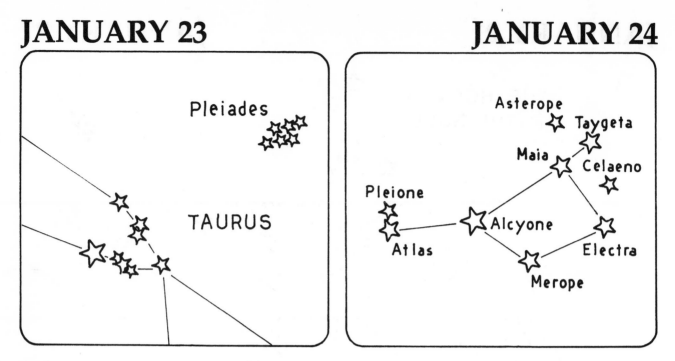

23rd: The *Pleiades* are a true cluster of stars, not an accidental alignment. Its members all lie about 400 light years from the earth. The brightest is Alcyone (*al-SIGH-oh-nee*), a star a thousand times more luminous than the sun. Photographs show the Pleiades embedded in wisps of gas which shine by light reflected from the stars. The cluster is young, having condensed from a great nebula only 20 million years ago. What are possibly glowing shreds of that nebula still cling to the stars.

24th: From very remote times the Pleiades have been a source of wonder, fascination, and delight. There is no other cluster of stars in the sky quite like it, a tight little "teacup" of stars that many people mistake for the Little Dipper. Legend refers to these stars as the seven daughters of Atlas (who held the world on his shoulders), or simply as the "Seven Sisters." There are actually hundreds of stars in this group, but only six can reliably be seen with the naked eye. Was another star of the cluster brighter in antiquity than at present? There is no way to know. For a pleasant story that accounts for the typical absence of the "seventh sister," see the description of the Big Dipper stars Mizar and Alcor on May 15. How many stars of the Pleiades can you see? Claims for naked-eye observations have ranged as high as 14–16. Nine stars are the most I have seen without the help of binoculars—on a spectacularly starry night!

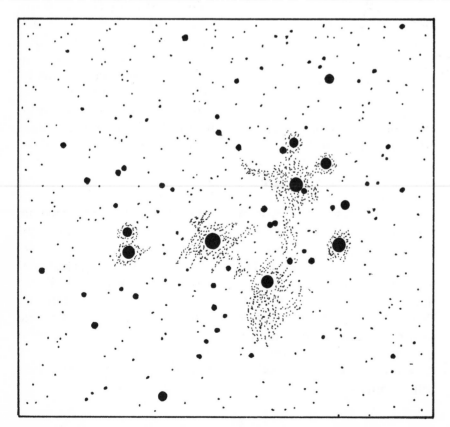

JANUARY 25

JANUARY 26

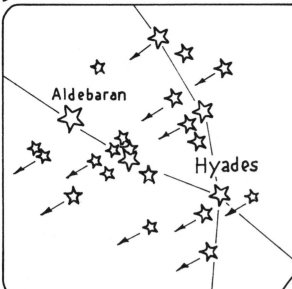

25th: Like the Pleiades, the *Hyades* are another striking example of an **open** (or **galactic**) **cluster.** These loose clusters of stars are typical of the spiral arms of our galaxy, and presumably represent groups of stars that evolved together from one of the great gassy nebulae that are common in the arms of the galaxy. Of the many stars in the Hyades cluster, the stars of the ''vee'' of the bull's face are the most easily observed. Aldebaran, although part of the ''vee'' and along the same general line of sight, is not a member of the group. The stars of the Hyades are moving away from us and toward the east at speeds of about 30 miles per second. Perspective makes them seem to be converging on a point in the sky near Betelgeuse in Orion, but it would take thousands of years for this so-called **proper motion** of the stars to become apparent to the eye. The fact that all of these stars move together confirms their common origin from a single nebula.

Crab Nebula

26th: Between the horns of the bull, but visible only with a telescope, is a remarkable object. The *Crab Nebula* is the shattered remnant of a star that blew itself apart in the year A.D. 1054—earth time. Since the star was many thousands of light years from earth, the explosion actually occurred thousands of years before it was observed here. This **supernova** explosion was recorded by Chinese and Japanese observers, but no European record is known. The explosion must have certainly been observed in Europe, because for weeks the supernova was the brightest starlike object in the sky. Almost a thousand years later, the cloud of debris is still expanding and is a beautiful object in telescopic photographs. Supernova explosions may be a typical fate for very massive stars. Only the collapsed core of a star might survive such a catastrophe.

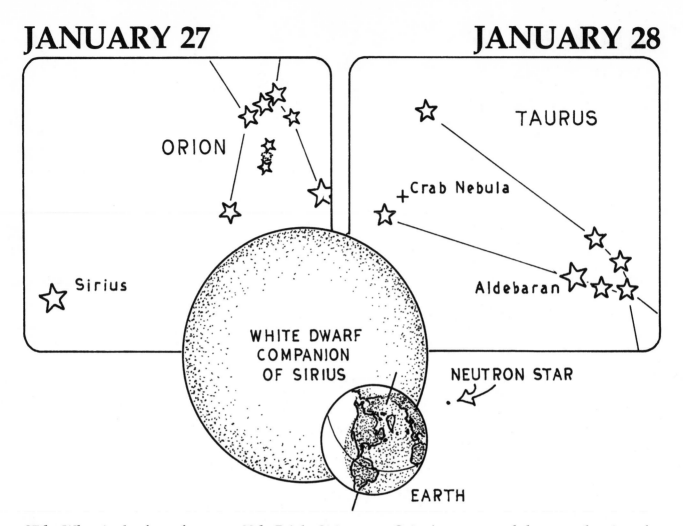

ORION

Sirius

WHITE DWARF
COMPANION
OF SIRIUS

EARTH

TAURUS

+ Crab Nebula

Aldebaran

NEUTRON STAR

27th: What is the fate of a star that has used up its nuclear fuel and temporarily swollen to become a **red giant,** or to the core of a star that has blown off its outer layers in a supernova explosion? Gravity again gets the upper hand and squeezes the remaining mass of the star to greater and greater densities. If the dying star has a mass about that of our sun or less, the squeeze will stop when the repulsive force between electrons in the matter of the star at last equals and resists the crush of gravity. At this point the star is about the size of the earth and has an almost unimaginable density. Such a star is white hot, but because of its small size not very bright. It is called a **white dwarf.**

28th: Bright Sirius, near Orion's foot, has a tiny white dwarf companion visible only with a telescope (see Feb. 11–14). Billions of years from now, after a period as a red giant, our sun too will spend the waning years of its life as an earth-sized dwarf. But if an exhausted and collapsing star has a mass greater than several times that of the sun, the repulsion force between electrons is not strong enough to resist the crush of gravity. The star continues to contract. Electrons are squeezed into protons to form neutrons. Finally, the repulsive force between the neutrons, which is stronger than that between the electrons, can stop the collapse. At this stage, a star originally larger than the sun has been

squeezed down to the size of a city, and is called a **neutron star.** A teaspoon of its matter would weigh a billion tons! The properties of matter in such a state of extreme density are almost too bizarre to be imagined. If the neutron star is rotating rapidly, it can radiate regular and rapid bursts of light, radio, and X-rays. It is then called a **pulsar.** The theoretical existence of neutron stars was predicted in the 1930s. The first pulsar was discovered in 1967. Since that time, many pulsars have been catalogued. There is a pulsar at the center of the Crab Nebula, the skeleton of the much more massive star that blew off its outer layers in that colossal explosion.

JANUARY 29

JANUARY 30

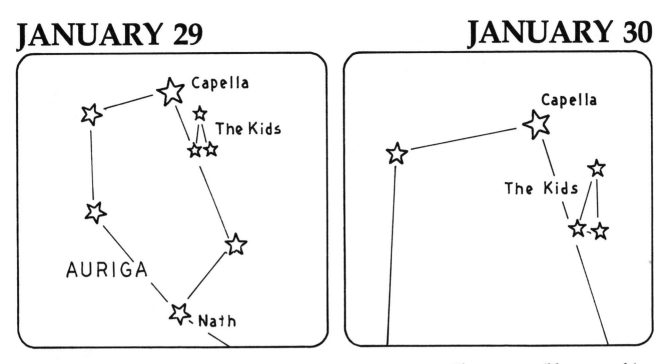

29th: Auriga (or-EYE-ga), the Charioteer, is not a constellation whose shape readily suggests the figure it is supposed to represent. Most people try to remember it simply as a pentagon. The most southerly star of the pentagon, Nath, is actually in Taurus and is the tip of the bull's horn. During the evening hours this month the constellation, including bright Capella, will be almost directly overhead.

30th: The dazzling yellow star near your **zenith** tonight is beautiful Capella. It is the fourth brightest star visible to northern observers (after Sirius, Arcturus, and Vega), and the sixth brightest star in the heavens. It is the most northerly of the first-magnitude stars. Capella is actually a **binary** or double-star system. Both components are about the same temperature as the sun, but many times larger and brighter.

They are possibly approaching the **red giant** stage of their evolution, or perhaps contracting from it. Capella means "she-goat," and the three little stars in a triangle just to the southwest of Capella are known as "the kids." Capella is a relatively nearby star, only 45 light years away. The distances of the kids from earth are 370, 1200, and 3000 light years—a long way from mother!

JANUARY 31

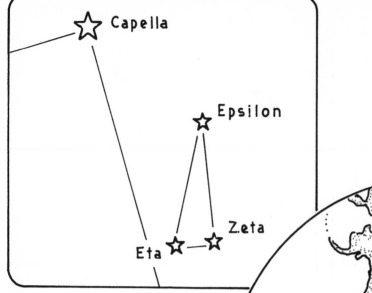

31st: Capella's kids are interesting in their own right, particularly Epsilon Aurigae. This is one of the most mysterious stars in the sky. Once every 25 years the light from Epsilon dims for a period of about 700 days. Some astronomers believe that this is because Epsilon is eclipsed by a huge invisible companion star, revolving around Epsilon and itself too cool and rarefied to emit much light. If this is so, Epsilon's companion would be the largest star known, about 3000 times bigger than the sun and about the size of the orbit of Saturn! Another possibility is that the companion is a more ordinary star, enshrouded and hidden by a great cloud of dust, possibly a solar system in the process of formation. More recent speculation has suggested that the mysterious unseen companion of Epsilon Aurigae might be a **black hole,** an exhausted very massive star collapsed to such incredible density that not even light can escape the pull of its gravity.

THE EDGE OF NIGHT
January 31

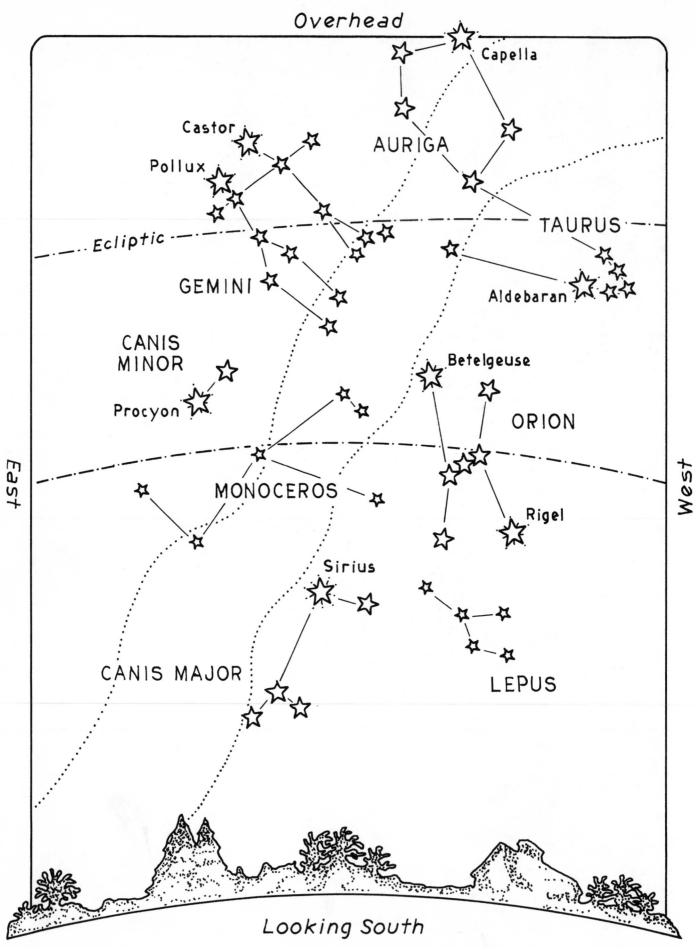

Overhead

Capella

AURIGA

Castor

Pollux

TAURUS

Ecliptic

Aldebaran

GEMINI

CANIS
MINOR

Betelgeuse

Procyon

ORION

East

West

MONOCEROS

Rigel

Sirius

CANIS MAJOR

LEPUS

Looking South

FEBRUARY 1

1st: Cold, clear winter nights are not always best for viewing the sky with a telescope. Often on those evenings when the sky seems wonderfully clear and the stars twinkle against a jet black background, there is turbulence in the atmosphere. It is this "shaky air" that makes the stars seem to twinkle. It can cause a telescopic image to shake and blur. But although winter skies do not offer ideal conditions for the telescopic observer, they are perfect for the naked-eye astronomer—if you can stand the cold. So put on your boots and mittens, wrap up in sweaters and scarves, and step outside. All winter long the evening sky is dominated by the brilliant star Sirius, the brightest of all the stars seen from the planet earth. Face south. Sirius will be staring you in the face, alone in its splendor. It is impossible to miss. The star will be about two handspans (arm extended) above the southern horizon, or about 30°. Sirius will be our guide star for the month of February. To the west (your right) are the familiar stars of Orion the Hunter. About one-and-a-half handspans above Sirius and a little toward the east is Procyon, the companion of Sirius in its journey across the winter sky. And high overhead, only a few finger breadths apart, are the Gemini twins, Castor and Pollux. These are the stars, with their associated constellations, which we shall study this month. They twinkle with a beauty unparalleled at any other time of the year, especially on a clear moonless night. But don't forget warm socks!

South

2nd: Can you find Orion on the drawing above? Then you will have no trouble discovering his three animal friends. Lepus the Hare darts to the left directly beneath the feet of the hunter, perhaps frightened by the bull that charges Orion from the west. Canis Major, the Big Dog, Orion's faithful hunting companion, follows fearlessly at his foot. Leaping excitedly near Orion's shoulder is Canis Minor, the Little Dog. You will have a hard time imagining the form of a dog among the stars of either constellation. The dog constellations are best learned (at first) as pairs of stars of unequal brightness. The Canis Major pair is a little brighter and a little more widely separated. Sirius dominates the scene. Since it is low in the sky, and therefore observed through a greater thickness of the earth's atmosphere, the star will have a noticeable tendency to "twinkle." To my eye, Sirius sparkles with a hint of blue, but you may see only a brilliant white. Only the planets, when they are in the sky, rival its brightness.

FEBRUARY 3

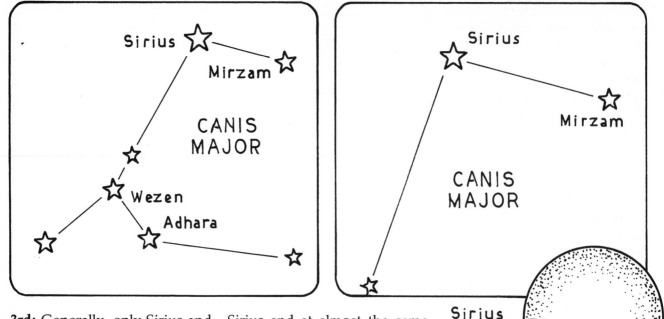

Sirius
Mirzam

CANIS
MAJOR

Wezen
Adhara

FEBRUARY 4

Sirius

Mirzam

CANIS
MAJOR

Sirius

Sun

3rd: Generally, only Sirius and its bright neighbor Mirzam are prominent among the stars of Canis Major. The other stars of the constellation are, for observers in northern latitudes, so close to the southern horizon that they are often lost in the haze. You will need a clear night and an open view to the south to see them. Mirzam (or sometimes Murzim) takes its name from the Arabic word for "announcer." It rises just before Sirius and at almost the same place on the eastern horizon, and "announces" the imminent arrival of that greater star. Mirzam is less bright than both Wezen *(WAY-zen)* and Adhara *(add-DAY-ra)*, but it is higher in the sky and usually more prominent. Actually, Adhara is the 22nd brightest star in the sky, almost as bright as more familiar Regulus, but is overshadowed by its brilliant neighbor Sirius.

4th: Sirius *(SEAR-ee-us)* takes its name from the Greek word for "scorching." It is more familiarly known as the Dog Star. No other star is as bright as this sparkling blue-white jewel of the southern skies. It is almost ten times brighter than first-magnitude stars such as Betelgeuse and Aldebaran. Sirius is bright for two reasons: it is one of our closest stellar neighbors in space, and it is a hot, luminous star in its own right. The surface temperature of Sirius is about 10,000°C, compared to the sun's temperature of about 6000°C. Of our one hundred nearest neighbors in space, Sirius is the hottest and most luminous. There are more distant stars that are intrinsically far brighter than Sirius, but around this part of the universe, Sirius is "the big boy on the block."

FEBRUARY 5

FEBRUARY 6

5th: Sirius is one of our nearest neighbors in the Milky Way Galaxy. Of all the stars that can be seen with the naked eye by most observers in North America and Europe, Sirius is the closest—excepting, of course, the sun. Only four other stars or multiple-star systems are nearer. Of these Alpha Centauri can be viewed only from southern latitudes (see Sept. 29), and three other stars are too faint to be seen without a telescope. Sirius is almost 9 light years away (53,000,000,000,000 miles). To get a sense of the scale of stellar distances, imagine the following: if the sun were the size of a Ping Pong ball, the earth would be a pinprick 13 feet away and Sirius would be a tennis ball 1400 miles away! A trip to Sirius in a spacecraft like the *Voyager*s, traveling at 50,000 miles per hour, would take more than one hundred thousand years. Given such vast distances, it may seem surprising that we can see the stars at all.

6th: The prominence of Sirius as the brightest of the stars has given it a special place in the myths of mankind. It has been an object of wonder and veneration in all cultures. The association of the star with the figure of a dog (or wolf) is surprisingly universal, and attests to the great antiquity of man's speculations concerning the stars. Sirius had special importance for the ancient Egyptians, who worshipped it as "the Nile Star." It rose in the east just before dawn on the first day of summer. Its appearance heralded the rising of the waters of the Nile River, an annual event of crucial significance to the agriculture-based civilization of that river valley. Several Egyptian temples were oriented so that the light of the rising star penetrated deep into their interiors. In other, more northerly cultures, the rising of Sirius with the sun meant that the scorching heat of July and August was at hand. In this respect, the influence of the star was considered malevolent.

Virgil writes of "the Dog Star, that burning constellation, when he brings drought and diseases on sickly mortals, rises and saddens the sky with inauspicious light." Even today, we speak of the heat of late summer as "dog days." A belief that the stars exert a significant influence upon human fortunes has been almost universal among our ancestors. The rise of modern science has reduced this belief to the level of quaint superstition. We no longer venerate or fear the star Sirius. But we are able—perhaps more able—to appreciate its great beauty as illuminator of winter skies.

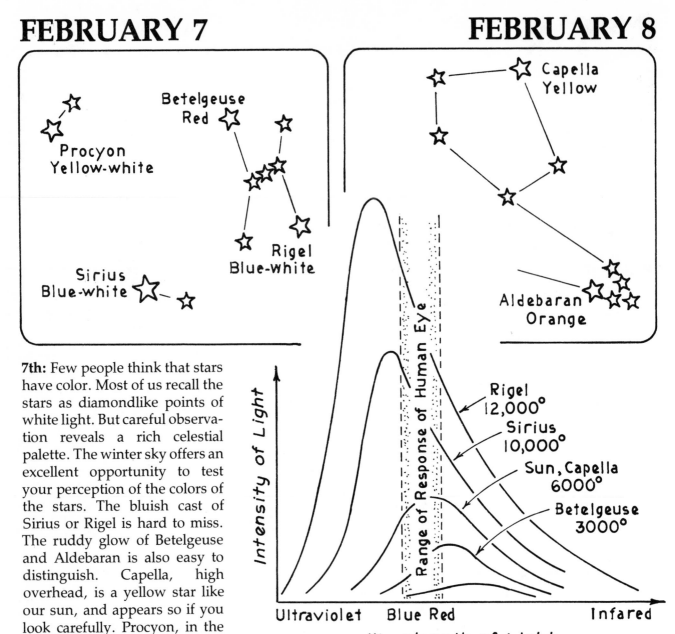

7th: Few people think that stars have color. Most of us recall the stars as diamondlike points of white light. But careful observation reveals a rich celestial palette. The winter sky offers an excellent opportunity to test your perception of the colors of the stars. The bluish cast of Sirius or Rigel is hard to miss. The ruddy glow of Betelgeuse and Aldebaran is also easy to distinguish. Capella, high overhead, is a yellow star like our sun, and appears so if you look carefully. Procyon, in the Little Dog, is a fierce yellow-white. The color of a star is determined by the temperature of the star's surface. The relationship is the same as for an iron poker in a fire. As the poker begins to heat up, it glows red hot, then orange, then yellow. If we continued to raise the temperature, the poker would appear white hot, or even white with a bluish cast. Match the color of the poker to the color of the star, and you have determined the temperature of the star.

8th: Light is an electromagnetic wave, and these waves can have differing wavelengths. Wavelength determines the color of light. Red light, for example, has a longer wavelength than blue light. Hot dense objects, like stars or pokers, emit a full rainbow of wavelengths called a *continuous spectrum*. The part of the spectrum to which the human eye is sensitive is called *visible light*. As the temperature of a luminous body increases, two things happen: the total brightness of the object increases, and the wavelength of peak intensity shifts toward the shorter wavelengths (or toward the blue end of the spectrum). It is the position in the spectrum of this peak intensity, relative to the visible part of the spectrum, that accounts for the color of the stars. The human eye is most sensitive to yellow light, possibly because we evolved near a yellow star!

A
F
Procyon

Betelgeuse
M
B

B
B
B

A
Rigel
B

Sirius
A
B

Castor
Pollux
A
K

K
F
M
M
A

Dark Line Spectrum of Sirius

Red

Yellow

Green

Blue

Violet

9th: If a prism is used to spread out the light from a star into its component colors, certain wavelengths (or colors) are found to be "missing" from an otherwise continuous rainbow of color. These dark lines, or missing wavelengths, are caused by the absorption of light by atoms or molecules in the cooler atmosphere of the star. Each kind of atom or molecule absorbs light at particular characteristic wave-lengths. Moreover, the degree to which certain atoms or molecules absorb light is strongly temperature-dependent. By examining the "dark lines" in a star's spectrum, the astronomer has another way to determine the star's temperature. The nature of the *dark-line spectrum* (or *absorption spectrum*) of a star defines its **spectral type,** and is designated by a letter code. Ranging from the hottest stars to the coolest, the letters designating spectral type are O-B-A-F-G-K-M. Students have long remembered this sequence with the aid of the somewhat sexist phrase, "Oh Be A Fine Girl, Kiss Me." The instrument the astronomer uses to examine a star's spectrum is called a *spectroscope.*

10th: The **spectral type** of some familiar stars, designated by letter, are shown on the maps above. Sirius is a hot blue-white A-type star. Betelgeuse is a cooler red M-type star. The sun is a G-type star, intermediate in the range of temperatures for stars. The surface temperature of the sun is about 6000°C.

Prism

Light from Star

FEBRUARY 11

FEBRUARY 12

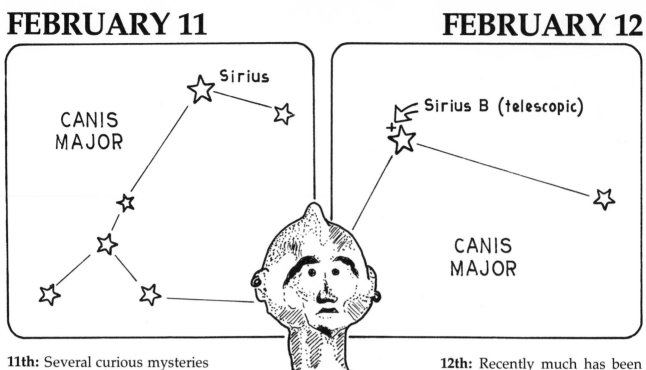

Dogon
Ancestor
Figure

11th: Several curious mysteries surround the star Sirius. The first arises from the fact that many ancient writers, including Cicero, Horace, Ptolemy, and Seneca, speak of Sirius as red or ruddy. To present-day observers, Sirius appears distinctly white, even bluish. Could the star have changed color since classical times? This seems unlikely. However, in the 19th century Sirius was discovered to have a tiny companion star, called Sirius B, a **white dwarf.** We know that old stars often pass through a red giant stage before beginning the final collapse toward becoming a white dwarf. Could Sirius B have been a **red giant** in the days of Cicero and Horace, outshining (because of its great size) the hotter blue-white star Sirius A? It would seem that such changes should take much longer than the 2000 years that have elapsed since the ancient writers commented on the star. Surely the sensitivity of our eyes has not changed. The mystery remains.

12th: Recently much has been made of the fact that the Dogon tribe of central Africa have a legend regarding an unseen companion of the sacred star Sirius. The legend relates that the companion star circles Sirius once every 50 years and is made of material heavier than any metal found on earth. These characteristics correspond remarkably well with our contemporary knowledge of the white dwarf Sirius B, knowledge that would be inaccessible to the nontelescopic, nonscientific observer. Another mystery! How did the Dogon acquire their knowledge of Sirius B? Pure coincidence? Was—as has been suggested in a popular book—the tribe visited by extraterrestrials who imparted to them special knowledge of the star? It seems far more likely that the Dogon were visited by a 19th-century missionary or traveler who passed along information concerning the recent discovery of Sirius B. The Dogon may have promptly incorporated this information into their religious myths!

FEBRUARY 13

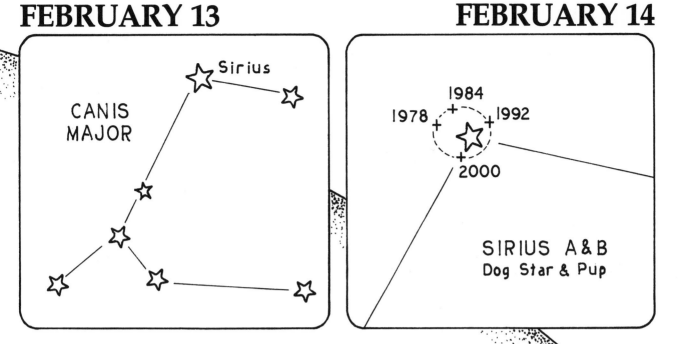

CANIS MAJOR

Sirius

FEBRUARY 14

SIRIUS A & B
Dog Star & Pup

1978   1984   1992
2000

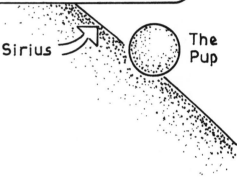

Sirius →   The Pup

13th: The existence of Sirius B was guessed in 1844 by the astronomer Friedrich Bessel, on the basis of tiny irregularities which he had observed in the slow **proper motion** (see Feb. 17) of Sirius across the sky. From wobbles in the motion of Sirius, he calculated that the unseen star orbited Sirius every 50 years, exerting its gentle gravitational tug on the larger body. In spite of a careful search, Sirius B was not observed until 1862, with what was then the largest refracting telescope in the world. The new star, called the Pup, was difficult to see in the glare of the far brighter Dog Star, Sirius.

The Sun

14th: From its motion about Sirius, the mass of the Pup can be calculated to be about equal to that of the sun (see Mar. 18). The spectrum of the Pup is that of an A-type star with a surface temperature of 9000°C, or about the same as Sirius. Both stars are white hot. Yet the Pup is only one ten-thousandth as bright as the "parent" star. How can this be? The answer is that Sirius B must be exceedingly small—not much larger, in fact, than the earth. With a mass equal to that of the sun (which has a diameter of nearly a million miles) packed into an earth-sized volume, the star must be astonishingly dense. A teaspoon of the mass of the Pup would weigh as much as an automobile! Such a star is known as a **white dwarf.** A white dwarf is thought to be the final stage in the life of a star such as the sun. Sirius B was one of the first white dwarf stars known, and it is still the best known of stars in that class. However, such stars are thought to be quite common. A billion years ago, the Pup was probably a star very similar to the sun, except for the presence of the larger more massive companion. Five billion years from now, our own sun may burn up the last dregs of its energy resources as a white dwarf star like the Pup. Does the Pup have planets? It is possible. The separation of Sirius A and B is over 20 times that of the earth and sun, enough to allow for planetary orbits about either star. But if life existed on a planet of the Pup, it is unlikely to have survived the death throes of that star—from sun-like star, to red giant, to slowly cooling white dwarf.

FEBRUARY 15

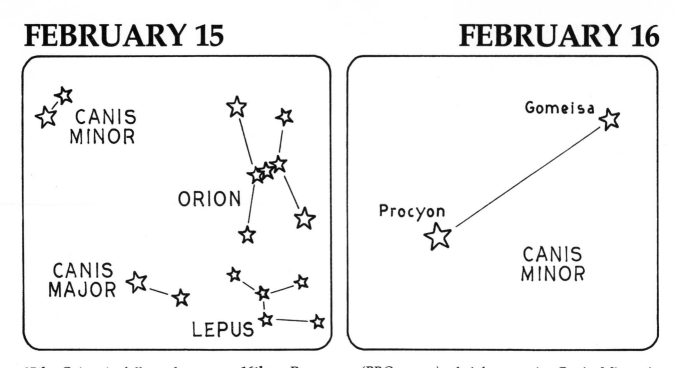

FEBRUARY 16

15th: Orion is followed across the heavens by three companions. The Big Dog (Canis Major), with bright Sirius, follows at the hunter's foot. Scampering between Orion's legs is Lepus the Hare, far enough south to be inconspicuous for northern observers. Jumping enthusiastically at Orion's shoulder, east of the great red star Betelgeuse, is the Little Dog (Canis Minor). In almost every respect, Canis Minor is a smaller version of Canis Major. Procyon is a little less bright than Sirius. It has a notable neighboring star, Gomeisa, a little less bright than Mirzam, the prominent neighbor of Sirius. The separation of the two stars in Canis Minor is slightly less than the separation of Sirius and Mirzam. Neither constellation looks much like a dog, but they look enough like each other to make finding them easy. A line extended eastward through the feet of Orion will lead to Canis Major. A line extended through Orion's shoulders will lead to Canis Minor.

16th: Procyon *(PRO-see-on)* means "before the dog." The star rises just before Sirius, although further north along the eastern horizon. The Egyptians must have watched eagerly for Procyon's appearance, anticipating the sacred Dog Star Sirius and the subsequent rising of the Nile. (Dogs figure prominently among the gods of the Egyptians. Illustrated below is the god Anubis, who had the head of a jackal.) The only other bright star in Canis Minor is Gomeisa *(go-MY-za),* which means, apparently, "bleary-eyed." The name, like that of many other stars, is of an Arabic origin. The Arabic astronomers were heirs to an even more ancient tradition. The actual origins of the names of most stars and constellations is lost in the mists of prehistory, and give us fascinating, if often obscure, insights into the minds of our remote ancestors.

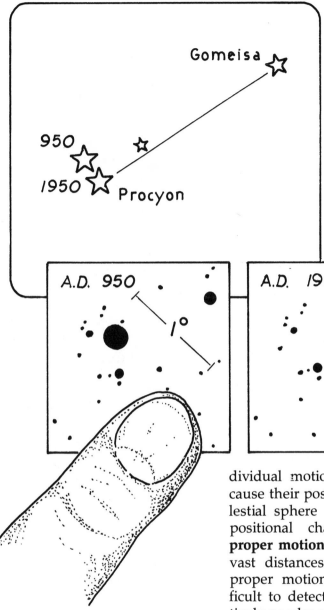

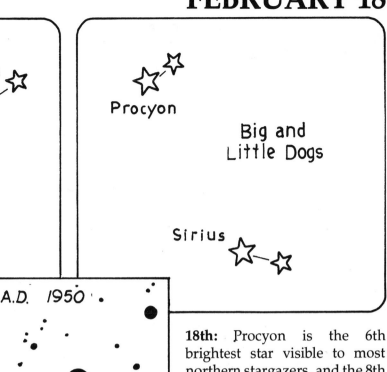

Big and
Little Dogs

18th: Procyon is the 6th brightest star visible to most northern stargazers, and the 8th brightest in all the heavens. It is our 14th closest neighbor among the stars, only 11 light years away. Of the stars familiar to northern observers, only Sirius is nearer. We have seen similarities in the constellations Canis Major and Canis Minor. The star Procyon is also a "little dog" when compared to Sirius. It is not quite so near or quite so bright as Sirius, but it is one of the nearest and brightest stars in the sky. The two stars have almost identical rates of proper motion. Like Sirius, Procyon has a white dwarf companion called Procyon B. The companion orbits the major star once every 40 years. The dwarf companion is hidden, except to large telescopes, in the glare of the larger star. The name given by Arabic astronomers to Procyon can be translated as "the northern Sirius." The similarity between the two stars is even greater than they imagined.

17th: We often refer to the stars as "fixed stars," a term used by the ancients to distinguish them from planets, or "wandering stars." Today we recognize, as the ancients did not, that stellar positions are not "fixed." All stars have some space motion relative to the sun. The stars in a galaxy are like bees in a swarm. The swarm moves as a whole, while within the swarm individual bees have motions proper to themselves. These in-dividual motions of the stars cause their positions on the ce-lestial sphere to change. This positional change is called **proper motion.** Because of the vast distances between stars, proper motions are very dif-ficult to detect. Only for rela-tively nearby stars is it possible to measure proper motion with any degree of accuracy. Since Procyon is near by, its proper motion is large compared to that of most other stars. In 1000 years, Procyon moves about one-half degree across the sky, or half the width of your little finger at arm's length. Only modern instrumental tech-niques, or a long and precise tradition of quantitative as-tronomical observation, could detect so small a change.

FEBRUARY 19

FEBRUARY 20

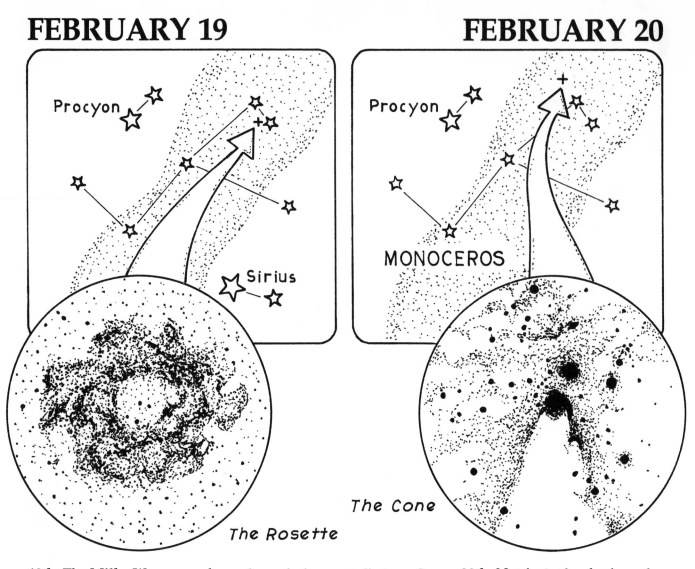

Procyon

Sirius

The Rosette

Procyon

MONOCEROS

The Cone

19th: The **Milky Way** passes between Sirius and Procyon. This band of pale light is actually the radiation of uncounted stars in the great flat disk of our spiral galaxy. Our star, the sun, is located about two-thirds of the way out from the center of this disk of stars. On starry winter nights, our view, is outward, away from the galaxy's center. In this direction the Milky Way is not so bright. On summer nights we look inward, toward the nucleus of the galaxy, and the Milky Way is more conspicuous. But if the night is clear, and you are far from city lights, look for the Milky Way tonight, arching from north to south, passing through the constellations Cassiopeia, Perseus, Auriga, and Orion before taking its headlong dive between the two Dogs. Here you will find the unremarkable constellation Monoceros, the Unicorn. The constellation is rather large in area, but contains no stars brighter than the 4th magnitude. You will need an excellent night to trace it. Monoceros contains several hidden treasures. One of these is the *Rosette Nebula*, a lovely wreath of glowing gas surrounding a cluster of young stars. Photographs show many dark tendrils or globes of gas which might be new stars in the process of formation.

20th: Not far in the sky from the Rosette is the *Cone Nebula*. Like the Rosette, it is only accessible to the casual stargazer by way of observatory photographs. But what spectacular photographs! The Cone appears as a dark pyramid of gas, part of a much more extensive region of nebulosity, illuminated from behind by the light of hot young stars. The tip of the cone is ablaze with stellar light. The aspect is that of a vast celestial candle, or dark divine tongue speaking stars. No photograph of any object in the heavens has, in my opinion, an equal power to excite awe and reverence in the mind of the viewer.

FEBRUARY 21

GEMINI

Castor
Pollux

Procyon

FEBRUARY 22

Castor
Pollux

Mebsuta

Wasat

Alhena

21st: Face south tonight and bend your head far back. High above you will see two bright stars about 5° apart (three finger breadths). Castor is the more northerly star, and if you look closely you will see that it is a little less bright than Pollux. In Greek myth, Castor and Pollux were the twin sons of the god Zeus and the mortal Leda. They were the brothers of Helen whose face "launched a thousand ships" and brought about the Trojan War. Castor was reputed to be a great horseman, and Pollux an expert boxer. Their names refer to those occupations. I have shown the twins as Greek athletes. In ancient times, Castor and Pollux were revered as protectors of sailors at sea. If you use the expression "by jimminy," you are swearing as the ancients did "by Gemini." With a little imagination it is not hard to trace the figures of the twins among the stars of the constellation, with the bodies of the two athletes arching toward Orion.

22nd: For the Chinese, the twin stars represented Yin and Yang, the opposed principles that give structure to the universe—light and dark, male and female, sky and earth, and so forth. A perfect balance of the two forces is the key to harmonious existence. But Pollux is somewhat brighter than Castor. Some authorities have argued that the two stars were more nearly equally bright in the past, and that one or the other has changed luminosity in recent times. If this is so, the perfect balance of Yin and Yang has been disturbed.

FEBRUARY 23

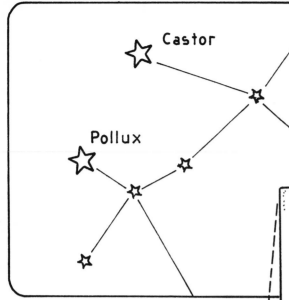

FEBRUARY 24

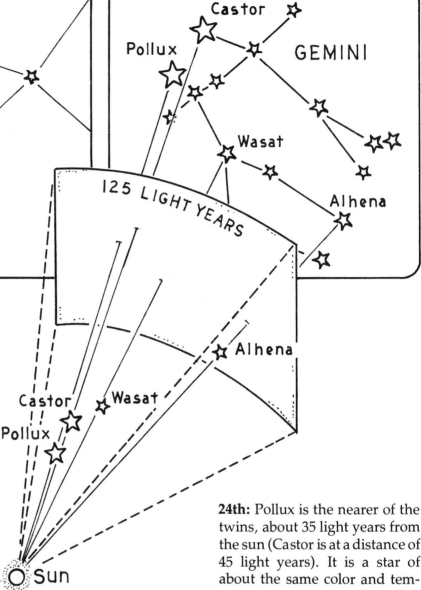

23rd: Castor is an exceedingly complex star system. What to the naked eye appears as a single bright star is revealed by a telescope to be a system of three stars (called Castor A, B, and C) in orbit about a common center of gravity. Further examination of the light of these stars with a spectroscope reveals that each of the three is itself a close pair of similar stars. Castor, then, is actually a system of six stars engaged in an intricate waltz about each other. The components of Castor A and B are hot A-type stars, rather larger than the sun, separated from each other by only a few million miles. Castor C consists of two tiny red stars, and is far removed from the dance of A and B. It is exciting to imagine the view from a planet in such a system. How much less exciting to live on the planet of a solitary star such as the sun!

24th: Pollux is the nearer of the twins, about 35 light years from the sun (Castor is at a distance of 45 light years). It is a star of about the same color and temperature as the sun, although it is considerably more luminous than the sun because of its size. Pollux is possibly a sunlike star on its way to becoming a **red giant.** It may, in fact, have increased in brightness in historic times. If this is so, and if Pollux continues to become more luminous as it swells in size, it may one day far outshine its nearby "twin." Future generations may be required to find a new name for the constellation.

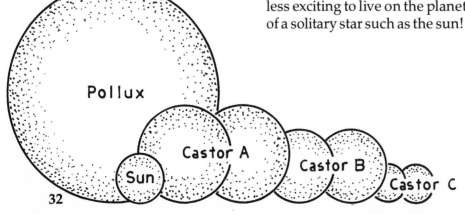

FEBRUARY 25

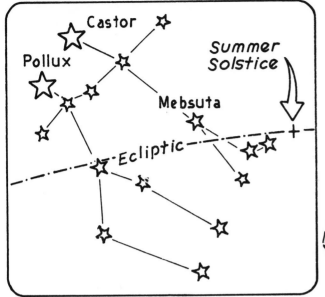

FEBRUARY 26

25th: Gemini is one of twelve constellations of the **zodiac.** The sun's yearly journey across the sky takes it through Gemini during June and July. The sun is in Gemini, just in front of Castor's "toe," on June 21, the day when it is furthest north in the sky. This special point in the sky, and the moment when the sun is there, is called the **summer solstice.** The solstice is therefore both a place (on the celestial sphere) and a time. As we have seen, the sun's path through the sky is called the **ecliptic.** Since the moon and the planets move in almost the same plane as the earth, we will always see those bodies somewhere in the sky near the ecliptic. On the evening of April 7, 1976, the planet Mars came this way and its tiny disk passed directly over the star Mebsuta. Quite by chance, I happened to observe this rare occultation of the star by Mars. Seen through binoculars, it was a beautiful event. Almost magically, two points of light briefly coalesced to one.

Herschel's 7-foot telescope

26th: Two planets of the solar system were discovered while in the constellation Gemini. On March 13, 1781, the great English astronomer William Herschel recognized the planet Uranus with the telescope illustrated here. The planet was at that time near the star Wasat. The discovery was quite unexpected. At first, Herschel thought he had found a comet. In fact, the object had been observed—and even plotted on sky maps—prior to Herschel's discovery, but was thought to be just another star. It was Herschel who recognized the "star" as a planet. Uranus was the first telescopic addition to the five naked-eye planets. Neptune followed. In February 1930, Pluto was discovered near the star Propus by Clyde Tombaugh at the Lowell Observatory in Arizona. The existence of still another planet had been suspected as the cause of tiny irregularities in the orbit of Uranus. Tombaugh found the planet after a year of patient study of photographic plates. Neptune, by the way, was in Capricorn at the time of its discovery in 1846.

FEBRUARY 27

Pollux

U Geminorum
+

Wasat

FEBRUARY 28

Pollux

Wasat

+
Eskimo
Nebula

Brightness 9 11 13 15

50 100 150 200 250

Days

27th: Like most constellations, Gemini harbors a number of intriguing objects visible only to the telescopic observer. One of these is the remarkable star U Geminorum. This star reminds us how unreliable some stars can be and how steadily our own wonderful sun continues to burn day by day, year by year. Every 100 days or so, U Geminorum suddenly flares up and for a day or so blazes a hundred times brighter than usual. Such stars are called *cataclysmic variables* or *miniature novae*. Why they have this erratic behavior is not well understood. It is known that U Geminorum is a close binary star system and the flare-up may be due to some kind of interaction between the two stars of the system.

28th: Another delightful telescopic object in Gemini has the dull technical designation NGC 2392 (NGC stands for *New General Catalogue*). Photographs of the object made with the most powerful telescopes suggest the more popular name *Eskimo*

Nebula. The object is of that class called **planetary nebulae,** spherical shells of gas ejected by aging stars and made to glow by energy radiated from the source star at the center of the nebula. As we look through these spherical shells, the thicker sides of the shells appear as roughly circular rings of gas. The objects were called "planetary" nebulae by 19th-century astronomers because the gassy circular blurs looked much like the fuzzy disks of planets. The most familiar planetary nebula in the Ring Nebula in Lyra.

34

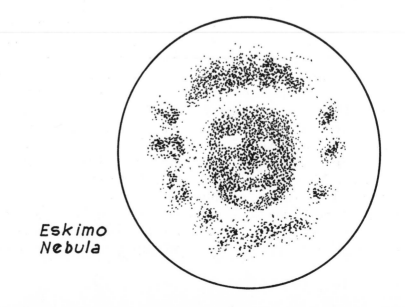

Eskimo
Nebula

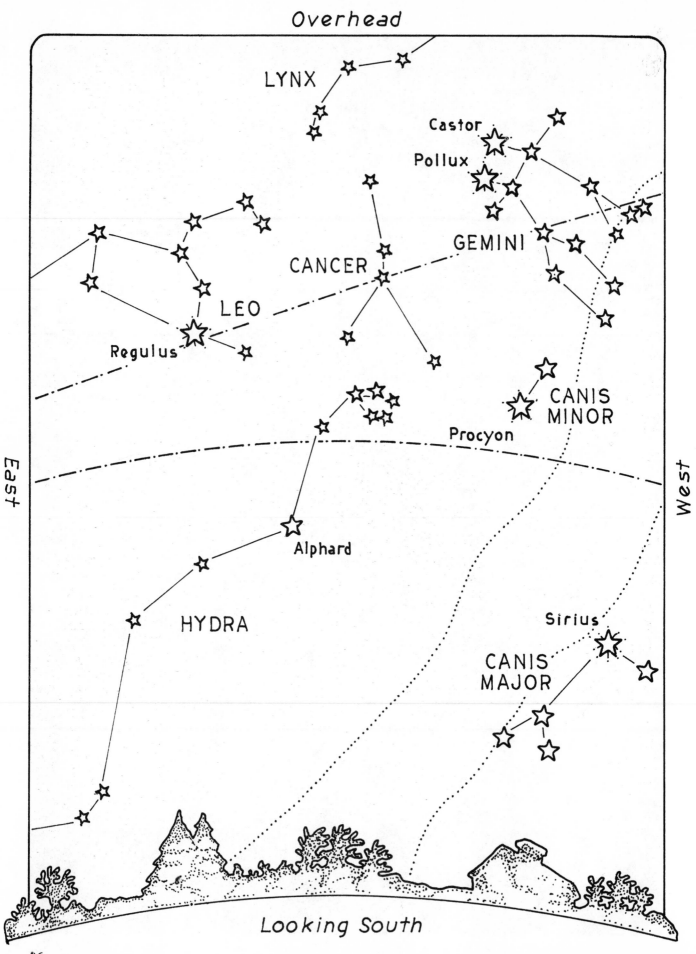

Overhead

LYNX

Castor

Pollux

CANCER

GEMINI

LEO

Regulus

CANIS
MINOR

Procyon

East

West

Alphard

HYDRA

Sirius

CANIS
MAJOR

Looking South

36

MARCH 1

1st: The red-winged blackbird is back. The pussy willows have put out their velvet buds. Everywhere there are signs of nature's rejuvenation. In the evening Arcturus is rising in the east, a sure sign of summer to come. The sky still has its bright stars. Orion and Gemini, the Big and Little Dogs, burn fiercely in the west. In the east, bright Regulus climbs on the ecliptic. But between the brilliant constellations of winter and the rising stars of summer is a dark void. The constellations we shall study in March—Cancer, Hydra, and the Lynx—are not readily recognized or easily learned. Except for Alphard in Hydra, no star of these constellations reaches the 2nd magnitude, and Alphard is low in the south. You will need a clear night to trace these feeble sky creatures—the crab, the water snake, and the lynx. But the faint stars we shall study are neatly bracketed by several of the most striking constellations in the sky, Gemini and Leo the Lion. You will have no difficulty knowing where to look for the objects of our study. The relatively starless void between Pollux and Regulus makes us appreciate all the more what Shakespeare called "night's candles." But why is the night dark at all? This curious question was first posed by Wilhelm Olbers in 1823. Olbers argued that in an infinite universe randomly filled with stars, the night sky should be ablaze with starlight, everywhere as bright as the very face of the sun. That the sky is not so lit is *"Olbers' Paradox."* Let us consider the matter further.

MARCH 2

2nd: Olbers reasoned as follows. Assume that the universe is infinite in extent, and that the stars are distributed with a uniform density throughout the volume of the universe. The intensity of a star's light should decrease with the distance of that star from the earth, since the energy radiated by the star is spread out over an increasingly large area. But the number of stars we encounter looking out through a wedge of space should increase with distance, and (it turns out) in mathematically exactly the inverse ratio at which the intensity of starlight diminishes. Weaker light at the greater distance, but more stars to contribute light. The two effects should just cancel. The conclusion: in an infinite uniform universe, the night sky should be ablaze with light. No matter which way you look, your line of sight should ultimately terminate on a star. But it is clear when we look out into the dark gulf of a March evening that the sky is not ablaze. How do we resolve the paradox? Perhaps the universe is not infinite, as Olbers assumed it was. A universe without limit or boundary is, after all, difficult to imagine. But at the same time, it is seemingly impossible to conceive of a universe which is not infinite; if limited, what lies beyond? Astronomers continue to debate the meaning of Olbers' Paradox. We shall come back to the question on the last day of the month.

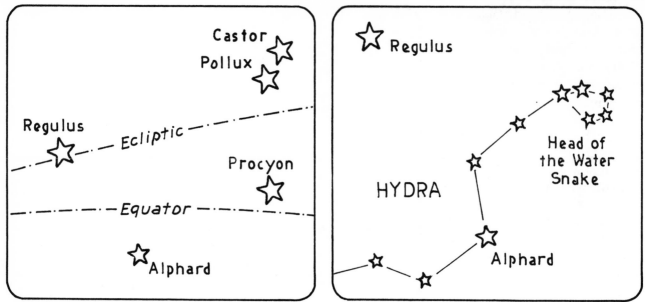

3rd: Let your eye follow the **equator** or the **ecliptic** across the dark gulf between Procyon and Pollux in the west and Regulus in the east. On a typical city or suburban night, the sky masked by haze and artificial light, and you might not see a single star to guide you along your way. These undistinguished parts of the sky make one wonder what lies between the stars. Is interstellar space empty? In the arms of the spiral galaxies the space between the stars is richly populated with gas and dust. Of course, "rich" is a relative term. Photographs of interstellar nebulae, such as the Rosette or the Cone, make those clouds seem thick indeed. But the average density of interstellar matter is far less than that of the earth's atmosphere, typically only a few atoms per cubic centimeter. And there is only one dust particle for every trillion atoms of gas. But even this small density of matter is sufficient to dim the light of stars in distant parts of the Milky Way Galaxy.

4th: Alphard, the "solitary one," stands alone in a part of the sky otherwise devoid of notable stars. But even in its isolation, Alphard is part of the great matter-transforming engine which is the **Galaxy.** The interstellar medium provides, in the dense nebulae, the material out of which new stars are born. In the furnace of a star's core, heavy elements are fused from light ones. The violent death throes of a star can throw off newly created elements to enrich the interstellar medium. The material of the Galaxy, then, is continuously recycled through the stars. The stars create "sticks and stones" from the thin stuff of interstellar space.

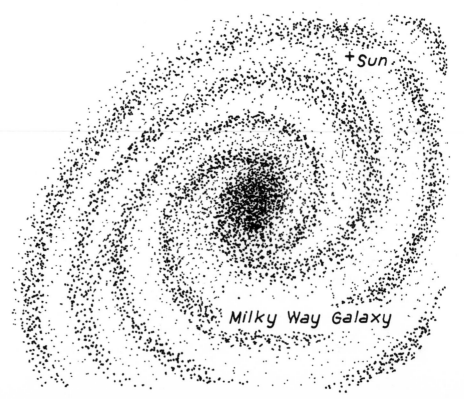

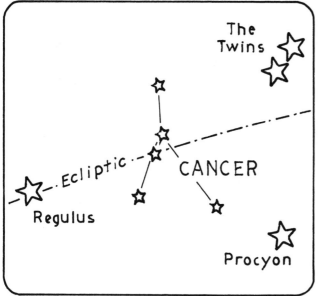

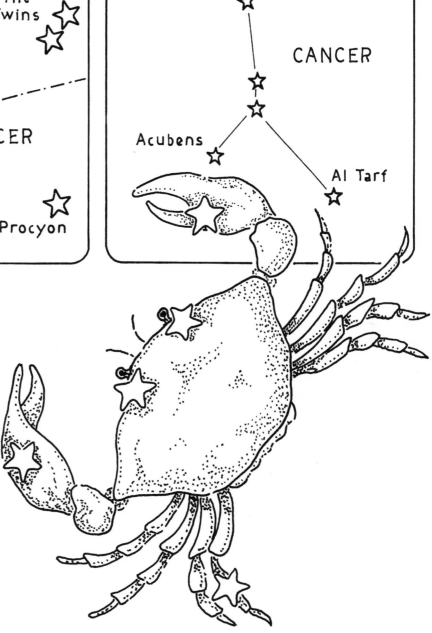

5th: There is no star in the constellation Cancer brighter than the 4th magnitude. You are not likely to see the crab (for that is what "Cancer" means) unless viewing conditions are excellent. City lights or haze will make it difficult to discern the constellation. Most old star maps show the crab as I have drawn it here: two stars for the eyes, two for the claws, and a fifth star at a back leg. Still, you will need a vivid imagination to see the figure of a crab among these faint points of light. None of Cancer's stars is particularly notable, and only a few have common names. Acubens, the southeastern star, means "the claws." Al Tarf, in the southwest, means "the end." Like many other stars, these have Arabic names. Cancer is an ancient constellation, one of 48 named constellations listed by the Alexandrian astronomer Claudius Ptolemy in his book *Almagest* compiled in the 2nd century A.D. Ptolemy came near the end of the great Greek tradition of astronomy.

6th: Cancer is one of the twelve constellations of the **zodiac.** The ecliptic passes close to the star which represents the southern eye of the crab. The sun comes this way in midsummer. On August 1 the sun stands almost exactly on the eye of the crab, as seen from earth against the background of the stars. Also, the moon passes this way each month, sometimes a little above, sometimes a little below the ecliptic and occasionally **occulting** stars of the constellation. If you see a bright "star" in Cancer, you can be virtually certain that it is a planet. I say "virtually" certain, because there is an *exceedingly* remote chance that you might be looking at a **nova.**

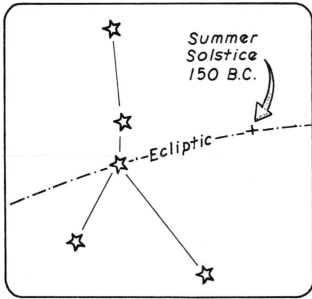

7th: 2000 years ago the sun was in Cancer when it reached the northernmost point of its journey across the sky. That point is called the **summer solstice.** The **Tropic of Cancer** took its name from the constellation. The tropic marks the latitude (23½° north) where the sun is directly overhead on midsummer day. The corresponding southern latitude, the **Tropic of Capricorn,** marks the sun's southernmost excursion, and takes its name from the constellation where the **winter solstice** stood in Greek times.

8th: The positions of the solstices have slowly shifted along the **ecliptic** over the years. The earth's axis, like that of a spinning top, makes a slow "wobble" under the stars. The wobble is called **precession.** One cycle of the wobble takes 26,000 years. This wobble causes the position of earth's poles and equator to change with respect to the stars (see May 22). As a result, the stars that coincide with the sun's most northerly excursion are different presently than in classical times. The summer solstice is now in the constellation Gemini. The wobble continues. In 1990, the position of the summer solstice will pass across the boundary of Gemini into the realm of Taurus the Bull (see Sept. 27). In 24,000 years the solstice will have passed through all twelve signs of the **zodiac,** back to the constellation that gave its name to Tropic of Cancer.

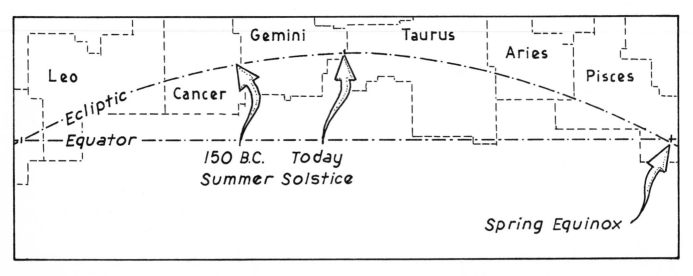

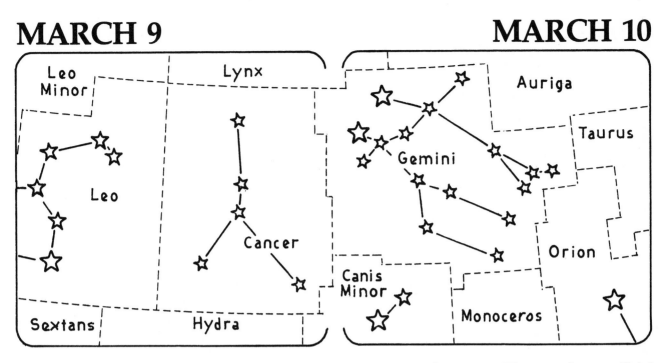

9th: Altogether there are 88 official constellations. A third of these are of little interest to the northern observer because they never rise above our southern horizon. Most of the northern constellations have had their present names since antiquity. Only a handful of the constellations in this book were not on Ptolemy's list of 48. The constellations, of course, are entirely arbitrary groups of stars, mostly derived from the imaginations of our remote ancestors. No new ones have been added since the 18th century.

10th: Cancer is certainly one of the least conspicuous constellations, but it is nicely bracketed by several of the most familiar groups in the sky. In older times, the boundaries of the constellations were ill-defined and represented only by informal curves established by long tradition. In the late 19th century, astronomers began to wish for more specific boundaries. The process of defining formal boundaries was completed in 1930 when the International Astronomical Union adopted the present configura-

tions. The modern official boundaries all follow north-south and east-west lines across the sky, but make a bewildering jigsaw puzzle of right-angle turns as they try to stay close to their traditional locations. Cancer is one of the more regular constellations, almost a rectangle. The only conspicuous northern constellation with perfectly rectangular borders is Canis Major. The modern constellations of the southern hemisphere tend to be more regular than those in the north.

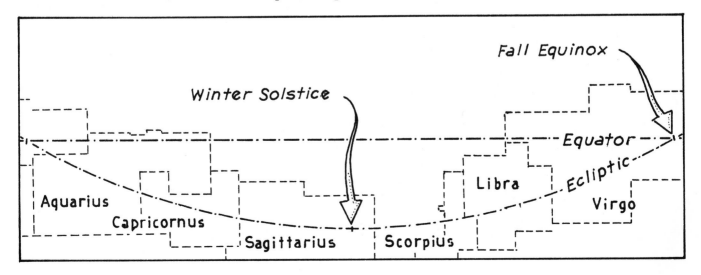

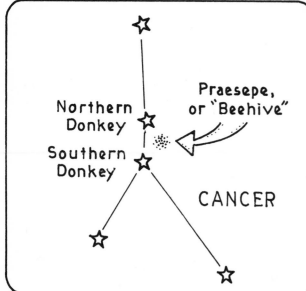

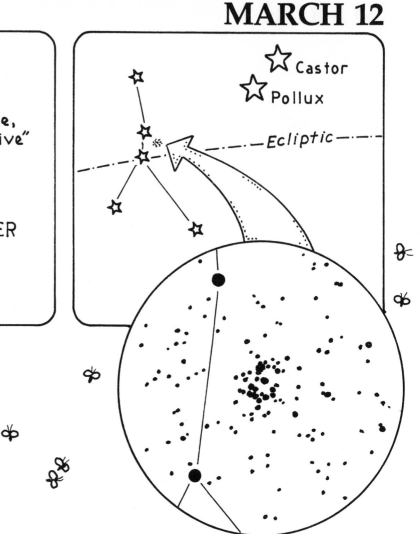

11th: If you are far from city lights and have a perfectly clear night, you may be surprised to see a small blur of light near the central stars of the crab. This nebulous object has been known since ancient times. In the days before pollution by smog and light, this tiny light-cloud must have been more prominent, more familiar, and more mysterious, too hazy to be a star, too small to be a fragment of the Milky Way.

12th: The little light-cloud in Cancer is called *Praesepe* (*pree-SEE-pee*), which means "manger." The two nearby stars, which I have illustrated as the eyes of the crab, are called "the donkeys." The donkeys, presumably, are eating hay from the manger, a rather picturesque image for a stellar blur. The most popular name for this wonderful object in Cancer is the *Beehive*. In 1610, when Galileo turned the first astronomical telescope (of his own construction) on this fuzzy spot, he was dazzled to view a cluster of tiny beelike stars. He counted almost 40. Today we can recognize almost 200 stars as members of the group. If you own binoculars, your instrument should show the stellar nature of this object. In fact, a good pair of binoculars is an excellent instrument for viewing the Beehive. The cluster is too large an object (about 1° across) to be visible all at once in the field of view of most telescopes. Since the Beehive is very near the ecliptic, you will sometimes find a planet visiting the hive. When a planet is in or near the hive, the view through binoculars is quite lovely, one of the delights the sky offers to even the casual stargazer.

MARCH 13

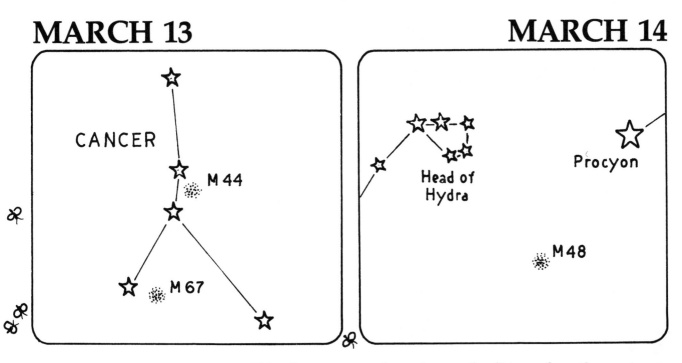

CANCER

M 44

M 67

MARCH 14

Head of
Hydra

Procyon

M 48

13th: The Beehive, after the conspicuous Pleiades, is the best-known example of an **open (or galactic) cluster.** It is sometimes referred to by its designation in **Messier's catalogue** of nebulae as M 44 (see Mar. 23). There is another open cluster, M 67, near the claw of the crab. Clusters such as these are common in the spiral arms of the Galaxy. They are typically scattered and irregular groups of hundreds of stars of various sizes and temperatures. Over 100 of the stars in the Beehive are brighter than the sun. A cluster might measure hundreds of light years across. All of the stars in an open cluster were probably born at about the same time from a single vast nebula of dust and gas. Often stars of these clusters are still surrounded by some of the nebulousity out of which they were born.

14th: Our sun may have been part of a cluster at its birth. If so, it has long since become separated from its siblings. In 5 billion years, the motion of the sun has carried it far from its place of birth. A cluster such as the Praesepe must be relatively young, with the "chicks" still in the nest. There is another rich open cluster of stars, M 48, near the head of Hydra the Water Snake, just south of Cancer and not far from Procyon. The cluster in Hydra has a diameter of 20 light years, or about five times

the distance from the sun to our nearest neighbor Alpha Centauri. It fills an area of the sky of about the size of the moon. The total light from the Hydra cluster is equivalent to the radiation of a thousand suns. You would need spectacular conditions to see it with the naked eye, but it is easily discerned with binoculars. These clusters lie at different distances from the earth. The Beehive is the nearest, at a distance of about 525 light years. M 48 and M 67 are about three times further away.

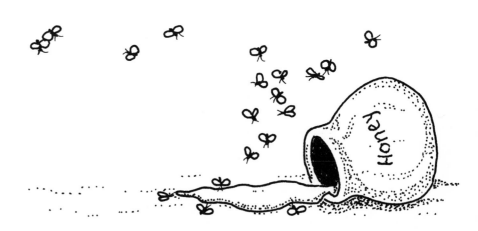

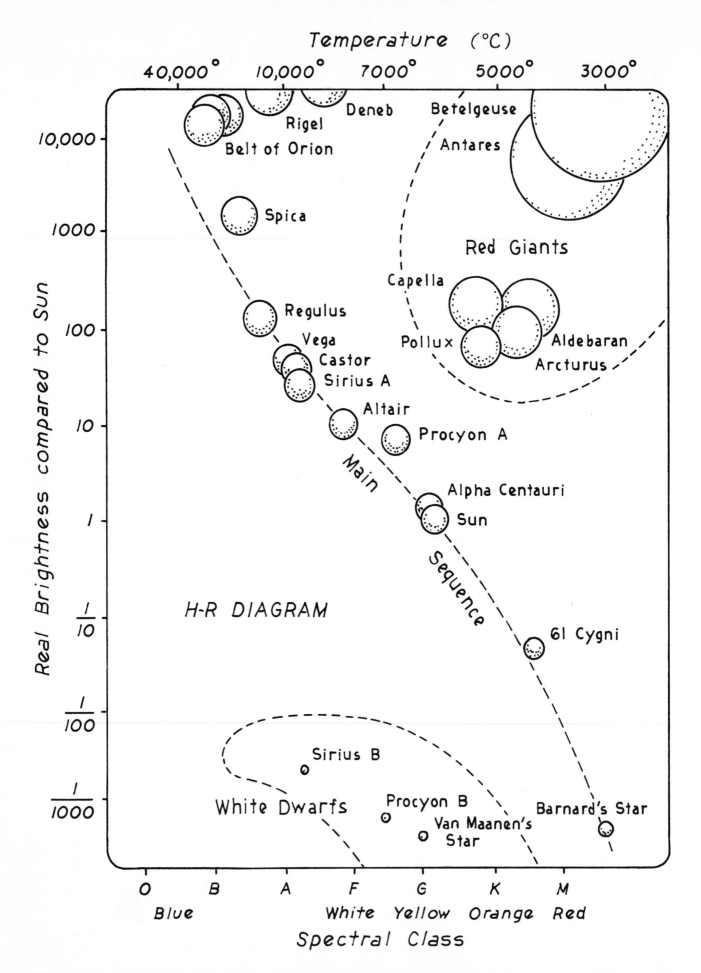

MARCH 15

15th: The stars of the Beehive are all about the same age, but have different colors and brightnesses. The color or spectral type of a star is determined by the star's surface temperature. Surface temperatures range from about 3000 to 40,000°C. If the star is cool (relatively speaking!), it is red; if hot, bluish. The real or absolute brightness of a star, or the total rate at which it radiates energy, is called **luminosity.** As a general rule, the luminosity of a star is directly related to its temperature. Typically—as one might expect—the hotter a star is, the brighter it is. If the real brightness of stars is plotted on a graph as a function of temperature, this typical pattern emerges and is called the **main sequence.** The sun is a main sequence star. So are Sirius, Procyon, Castor, and Regulus. The kind of graph we are talking about was developed by the astronomers Ejnar Hertzsprung and Henry Norris Russell in the early part of the 20th century. It is called an **H-R diagram.** The graph is a concise and elegant way to summarize the properties of stars. It turns out that some stars do not follow the general rule relating temperature and luminosity. They do not fall on the main sequence when plotted on an H-R diagram. Aldebaran and Betelgeuse, for example, are cool red M-type stars, yet we know that they are intrinsically very bright. They must, therefore, be much larger than typical red main sequence stars. We call them **red giants.** The great size of red giants makes them conspicuously bright even at great

MARCH 16

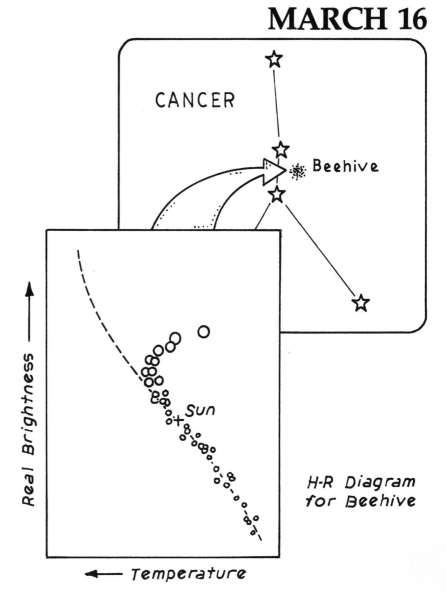

H-R Diagram for Beehive

distances from the earth. Familiar red giant stars are shown on the H-R diagram at left. Some other stars, such as the companions of Sirius and Procyon, are white hot and yet not very bright. They must be considerably smaller than the typical main sequence white star. They are called **white dwarfs.** White dwarfs are generally less than a thousandth as bright as the sun, and can only be observed if they are nearby. All of the visible stars have their own place on the H-R diagram.

16th: An **H-R diagram** poses intriguing questions regarding the lives of stars. If an H-R diagram is constructed for stars of the Beehive, a curious pattern emerges. Most of the stars fall on the main sequence. But the brighter stars of the cluster lie off the main sequence toward the red giant part of the diagram. There are no hot blue stars at the top of the main sequence. What's going on? This is a puzzle we shall answer over the next few starry nights. The sun is shown on the diagram as a convenient reference.

MARCH 17

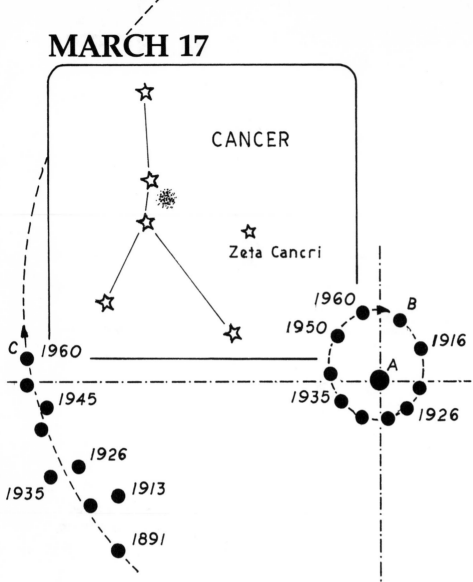

CANCER

Zeta Cancri

MARCH 18

18th: Zeta Cancri is an interesting multiple system consisting of three visible stars (A, B, and C) and an unseen white dwarf. The recent motions of these stars (with respect to A) are shown on this page. An analysis of the motions shows that all four components have masses similar to the sun's. It is the *mass* of a star that determines how brightly it burns—and its place on the main sequence. Unless the mass of a newly forming star is at least 1/100th that of the sun, its core will never reach temperatures at which nuclear fusion can occur. The object will not become a star at all, only a dark planetlike body. If Jupiter's mass were ten times greater, it would burn as a tiny red dwarf star and our sun would be part of a binary system. At the other end of the scale, stars with masses greater than 100 times that of the sun are rare. We shall shortly see one reason why.

17th: Binary star systems give astronomers an opportunity to determine the masses of stars. Binary stars are held in orbit about one another by the attraction of gravity. The force of gravity depends upon the mass of the attracting bodies. If a binary star is close enough to us that its distance can be independently measured (by **parallax;** see June 5–8), and if the orbital motion of the component stars can be observed over a period of time, then the masses can be calculated from known laws of gravity and dynamics. The mass of the sun can be calculated from its gravitational effect on the planets.

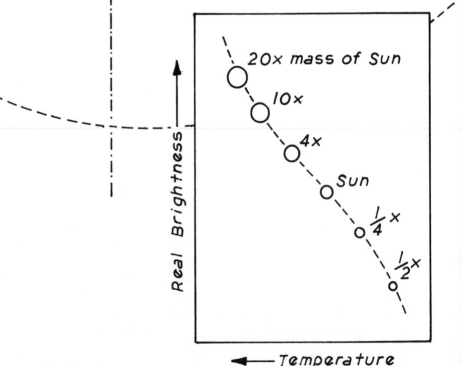

20x mass of Sun

10x

4x

Sun

¼ x

½ x

Real Brightness

← Temperature

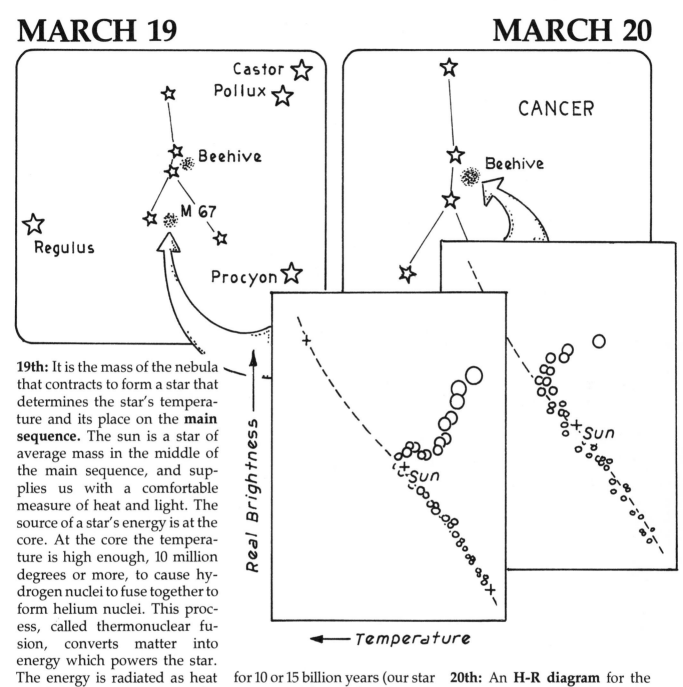

CANCER

Castor
Pollux
Beehive
M 67
Regulus
Procyon

Beehive

Real Brightness

Temperature

19th: It is the mass of the nebula that contracts to form a star that determines the star's temperature and its place on the **main sequence.** The sun is a star of average mass in the middle of the main sequence, and supplies us with a comfortable measure of heat and light. The source of a star's energy is at the core. At the core the temperature is high enough, 10 million degrees or more, to cause hydrogen nuclei to fuse together to form helium nuclei. This process, called thermonuclear fusion, converts matter into energy which powers the star. The energy is radiated as heat and light. For how long can the process continue? That depends on two things: the amount of hydrogen fuel available at the core, and the rate at which it "burns." More massive stars have more fuel, but they also have hotter cores and burn their fuel at a much quicker rate than the less massive stars. Calculations show that a mid-sized star like the sun can probably burn steadily on the main sequence

for 10 or 15 billion years (our star is now middle-aged). A hot star such as Spica might burn for only 10 million years before its resources are depleted. A cool, slow-burning red dwarf, at the bottom of the main sequence, can survive happily, the pressure of fusion balanced against gravity, for hundreds of billions of years. When a star exhausts its hydrogen resources, it swells to become a **red giant** and leaves the main sequence.

20th: An **H-R diagram** for the stars in the cluster M 67 shows that all stars more massive than the sun have burned up their hydrogen resources and left the main sequence. This suggests an age for the cluster of about 10 billion years. The "turnoff point," where the stars of a cluster leave the **main sequence,** tells us the age of the cluster. The Beehive is younger than M 67. It is only 1 or 2 billion years old.

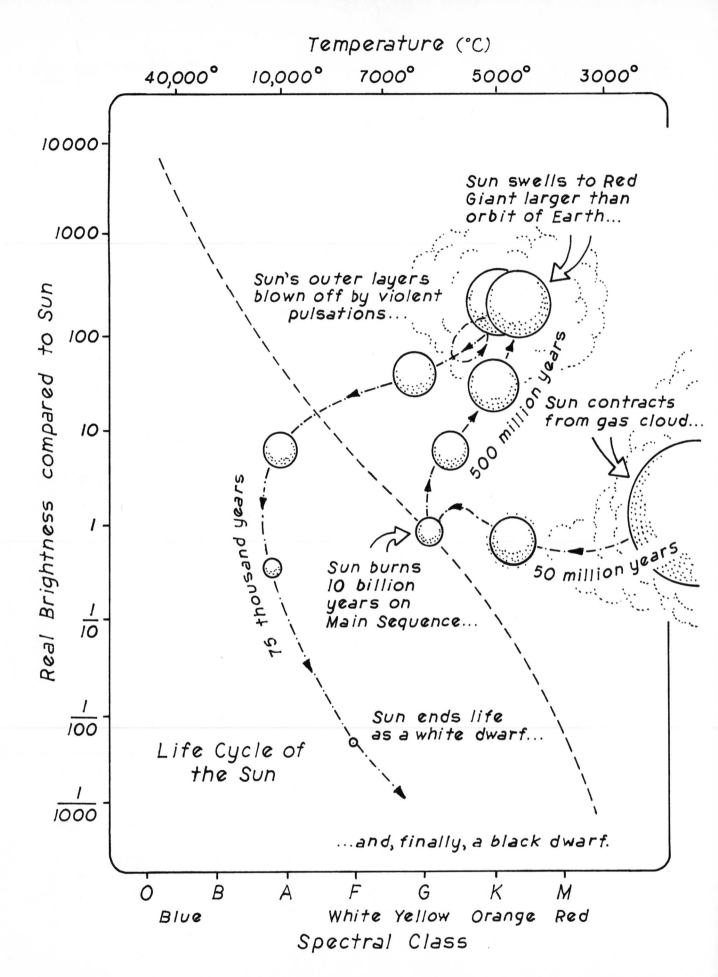

Temperature (°C)

Life Cycle of the Sun

Sun swells to Red Giant larger than orbit of Earth...

Sun's outer layers blown off by violent pulsations...

Sun contracts from gas cloud...

500 million years

50 million years

Sun burns 10 billion years on Main Sequence...

75 thousand years

Sun ends life as a white dwarf...

...and, finally, a black dwarf.

Real Brightness compared to Sun

O   B   A   F   G   K   M

Blue   White Yellow Orange Red

Spectral Class

MARCH 21

MARCH 22

21st: The first day of spring! The sun climbs back across the celestial equator into the northern sky. Each day we feel more of its heat, warming and refreshing the earth. The shadows grow shorter. The soil, long dormant, begins to stir. Crocuses push up their thin green fingers to clutch the light. Citizens of the earth's northern hemisphere lean toward the sun and welcome its flood of life. Briefly, but sharply, our attention is focused on one star above all others, a yellow middle-sized G-type star just 93 million miles away. How long will the sun continue its steady outpouring of heat and light? How long will the cycle of the seasons repeat itself with such welcome regularity? Calculations of the lifetimes of stars are exceedingly difficult, and are based on some of the most complex theories of physics. They require making intelligent assumptions about the interiors of stars. And they require a knowledge of atomic nuclei that has been gleaned over almost a century of patient investigation. These calculations suggest that a star the size of our sun can burn steadily for at least 10 bil-

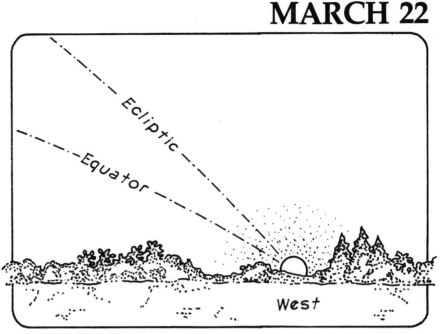

lion years. A study of the radioactivity of rocks and meteorites indicates that the age of the solar system is 4.6 billion years. If this is so, then our star is halfway through its life cycle—and we can safely expect another 5 billion verdant springs. As nearly as we can guess, the life cycle of our sun will be something like that shown on the drawing at the left. Four-and-a-half billion years ago, the star condensed from an interstellar nebula. At the point in its contraction when sufficient nuclear energy was triggered at the core to counteract the squeeze of gravity, the sun reached the **main sequence** and began its "normal" life. When the sun's internal energy resources are depleted, it will go through a period of instability (not well understood), briefly becoming a **red giant.** Then gravity will get the upper hand, and the sun will shrink to become a **white dwarf,** and—finally—a dead black star.

22nd: Tonight the sun goes down exactly due west. But even as it drops below the horizon it creeps northward along the ecliptic and crosses the sky's equator. Each night from now on it will stay in the sky a little longer and set a little further to the north. Summer is not far away! What will become of the earth when our star goes through its death throes in another 5 billion years or so? The prospect is not encouraging. First, the sun will swell until it becomes a huge red star filling the entire sky. The earth will be incinerated and life extinguished. Outer layers of the sun may be blown off to form a planetary nebula, but no one will be around to admire it from the inside. Finally, as the skeleton star shrinks to a white dwarf and slowly cools, it will radiate the last dregs of heat and light. If the earth has survived the period of violence which will accompany our star's death, it will end its days as a frozen cinder.

MARCH 23

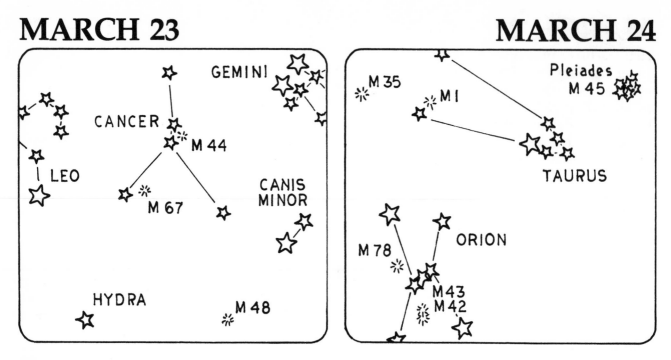

MARCH 24

23rd: Let us take a last look at the Beehive, that curious "fuzzy spot" in Cancer that can be seen with the naked eye on a perfect night. The Beehive is also known by its "M" designation, M 44. It is the 44th "fuzzy spot" in a catalogue of nebulous objects compiled by the French astronomer Charles Messier (*mess-YAY*) who lived from 1730 to 1817. There are 110 objects listed in **Messier's catalogue.**

24th: Charles Messier was a comet hunter. Comets are members of our solar system, in orbit about the sun. Except for a few of the very brightest, comets appear in the sky as faint blurs of light visible only with optical aid. A "blur" can be recognized as a comet by the way it moves. As an aid to comet hunters, Messier compiled a list of other celestial blurs—stationary blurs—that might be temporarily mistaken for comets. He

listed over a hundred. One of the objects he catalogued, the Pleiades, is easily recognized as a cluster of stars even with the naked eye. Others, like the Beehive, could be resolved into stars with the help of a telescope. But most of the objects in Messier's catalogue were simply fuzzy patches of light against a black sky—he had no idea what they were. We still refer to most of these objects by their Messier numbers.

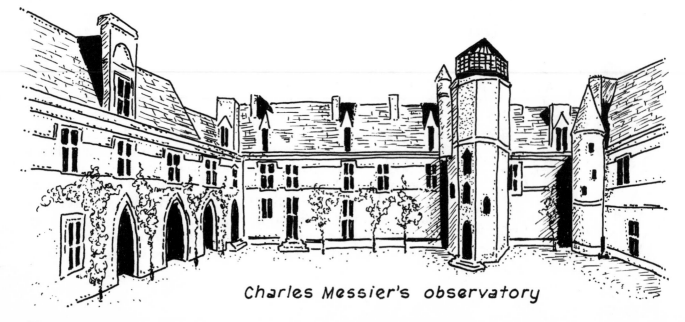

Charles Messier's observatory

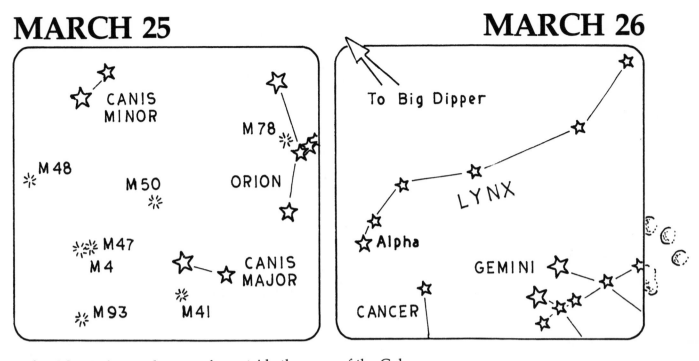

25th: Messier's catalogue of "fuzzy spots" turns out to contain objects of importance beyond anything Messier could have dreamed. The list remains valuable today, and not just for comet hunters. The blurs in Messier's catalogue consist of several different kinds of objects. Some, like the Beehive, and most of the other objects in this part of the sky, are **open clusters** of stars in the spiral arms of the Galaxy. These clusters typically contain hundreds of stars, and over a thousand such clusters have been recognized. A second type of object on Messier's list, such as M 13 in Hercules, consists of tight globe-shaped clusters of tens or hundreds of thousands of stars. These **globular clusters** reside

outside the arms of the Galaxy, but are bound to the Galaxy by gravity and form a kind of halo around it. A third kind of fuzzy spot with M numbers are the gaseous nebulae. Some of these are the debris of stars that have blown themselves apart in nova or supernova explosions, such as M 1, the famous Crab Nebula in Taurus. Others are the ex-

tended clouds of dust and gas out of which new stars are born, such as M 42-43, the Great Orion Nebula. It required telescopes larger than that used by Charles Messier in the 18th century to reveal the true natures of these remarkable objects.

26th: Here is another of the faint constellations in the great black gulf of the March sky. You will need an excellent starry night to see any of the stars of the Lynx, and even if you find them you will have a hard time visualizing the catlike little animal. I like to imagine the faint line of stars that makes up the constellation as merely the tracks of the Lynx as it stalks its prey across the sky. Alpha Lyncis is the brightest star in the constellation (magnitude 3.3), and the one you are most likely to see. Look for it in the dark and empty void just north of Cancer, between the easily recognized Gemini twins and the Big Dipper.

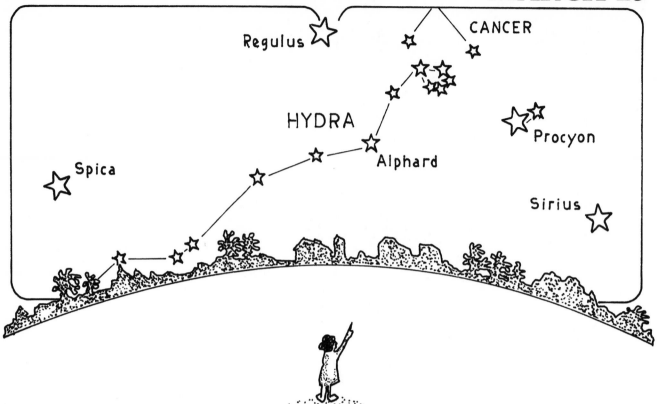

27th: Hydra is sometimes called the "many-headed monster," or sometimes the "water snake." Like other constellations we have studied during the not-so-starry nights of March, Hydra is not particularly conspicuous. To make matters worse, it is low in the southern sky and therefore a difficult object for most northern observers to see. Hydra does have one exceptional claim to fame—it is the largest of all the constellations in the heavens. It is the largest in area and in length. The long writhing body of this water monster stretches one-quarter of the way around the sky. The constellation has only one fairly bright star, Alphard. If you have a clear view of the southern horizon, you should have no trouble finding Alphard; it has little competition in this part of the sky.

28th: There are several myths surrounding the constellation Hydra. The most famous of these involves the great hero Hercules, the son of Zeus and the mortal woman Alcmene. (The illegitimate offspring of Zeus are not at all uncommon in the sky!) Hera, the divine wife of Zeus, bore no great affection for her husband's unlawful progeny. She cast a spell of madness over Hercules, and while out of his senses the unfortunate man killed his own children. When he recovered from the spell, Hercules undertook twelve terrible labors in restitution for his crime. The second of these twelve tasks required Hercules to slay the many-headed water monster Hydra. Each time Hercules hacked a head from the monster, two grew back. For a while it seemed to be a losing battle.

Then Hercules thought to take a torch and scorch the stump of each neck as he cut off a head. This did the trick. He buried the last and immortal head of the monster beneath a large stone. The Hydra was aided in this colossal battle by the giant crab Cancer, not far away in the sky. The great hero himself can be found among the constellations, and we shall take a look at him in July. Several of the twelve labors of Hercules have some reminder among the stars. There may, in fact, be some remote connection between the twelve labors and the twelve signs of the zodiac. The constellations provide a delightful textbook of the Greek myths. We shall touch on only a few myths during our 365 starry nights. You may want to read Gallant's book (see "Sources and Resources").

MARCH 31

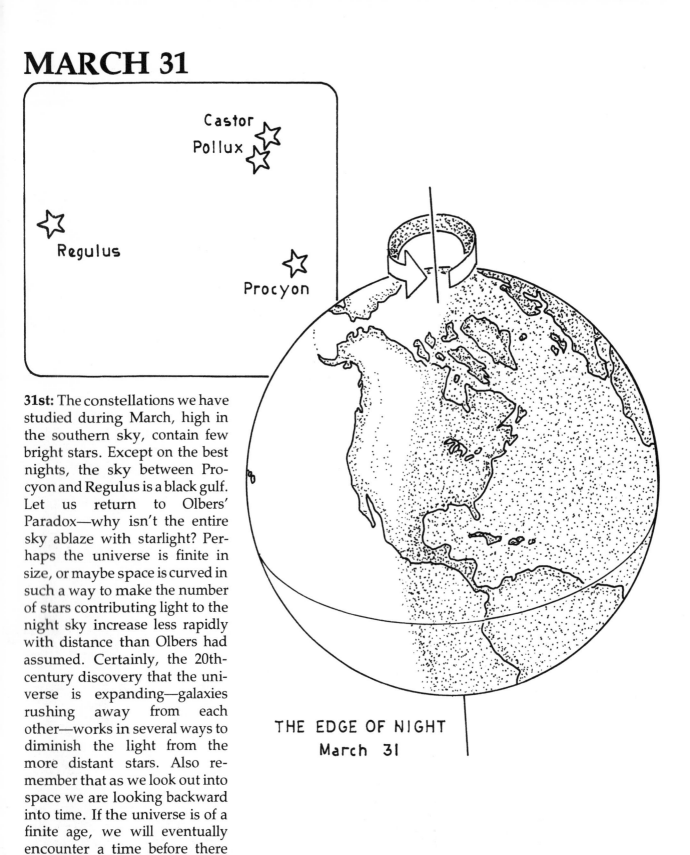

Castor
Pollux
Regulus
Procyon

THE EDGE OF NIGHT
March 31

31st: The constellations we have studied during March, high in the southern sky, contain few bright stars. Except on the best nights, the sky between Procyon and Regulus is a black gulf. Let us return to Olbers' Paradox—why isn't the entire sky ablaze with starlight? Perhaps the universe is finite in size, or maybe space is curved in such a way to make the number of stars contributing light to the night sky increase less rapidly with distance than Olbers had assumed. Certainly, the 20th-century discovery that the universe is expanding—galaxies rushing away from each other—works in several ways to diminish the light from the more distant stars. Also remember that as we look out into space we are looking backward into time. If the universe is of a finite age, we will eventually encounter a time before there were stars. Olbers' simple speculation of the darkness of the night sky leads us into the thorniest questions of modern astronomy.

29th: The head of Hydra is a nice little group of stars worth looking for on a clear night. It lies almost exactly halfway between Procyon in Canis Minor and Regulus in Leo. But the stars in the head of the water monster cannot compete with Alphard (*AL-fard*), the only bright star in this part of the sky. Alphard means "solitary one," an apt name. The star is also called Cor Hydrae, the "Hydra's heart." The heart of the monster is a 2nd-magnitude star. It is 95 light years distant from the earth and is an orange giant star intrinsically brighter than a hundred suns.

30th: Low on Hydra's back, near the bright star Spica in Virgo, are two minor constellations, Crater the Cup and Corvus the Crow. They are best viewed in April or May when they are highest above the southern horizon, but since they are related to Hydra in myth I will mention them now. According to the story, Apollo sent his pet crow down to earth to fetch a cup of fresh water. The crow lingered at the spring waiting for figs to ripen on a nearby tree, and when the figs were at last ripe, had himself a nice feast. To cover his delay, the crow plucked out a water snake

from the spring and brough home with the cup of wa blaming the snake for the del Apollo was not fooled, a flung the whole lot—crow, cu and snake—into the sky. Th you will see them today, t little rectangles of stars on back of the meandering snal Remember to look for them la in the spring.

Hydra, Crater, and Corvus
(after a 17th-century star
map by Andreas Cellerius)

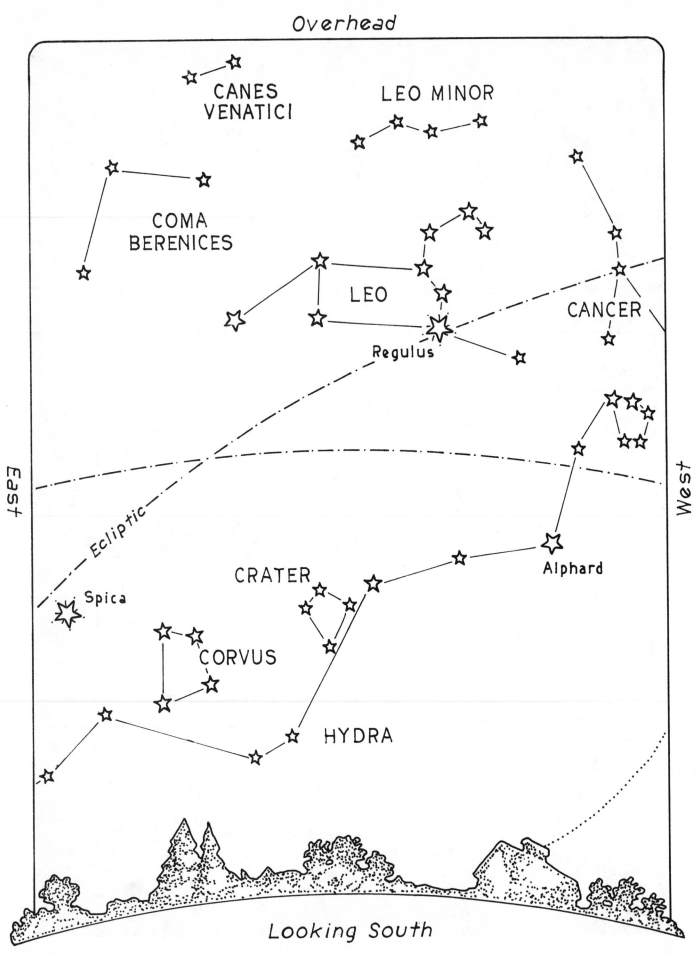

Overhead

CANES
VENATICI

LEO MINOR

COMA
BERENICES

LEO

Regulus

CANCER

East

Ecliptic

West

Spica

Alphard

CRATER

CORVUS

HYDRA

Looking South

APRIL 1

1st: In sandy soil at the side of the road the fiddlehead ferns unroll their sun-catching fronds. It is truly spring, and the sun moves northward on its long arc. Every second the sun converts 657 million tons of hydrogen into 653 million tons of helium by a process called nuclear fusion. The missing 4 million tons of mass are converted into energy and hurled into space as heat and light. The earth intercepts only about one two-billionth of this energy, or about 4 pounds worth of the vanished matter. The sun never misses so tiny a fraction of its huge bulk, but to the earth that 4 pounds worth of energy every second is the difference between day and night. It is also the difference between winter and summer, between death and life. In its journey around the sun the earth leans into its curve like a sailor bracing against the wind. Now in April we begin to lean toward the sun. The sun climbs higher and higher into the sky. Its rays hit the earth's surface more directly here in the northern hemisphere and the earth responds. In the summer, about a millionth of an ounce of the sun's mass falls every second onto the college campus where I teach; in the winter less than half as much. A fraction of a millionth of an ounce of the sun's depleted mass is all it takes to tip the balance of the season back toward winter or forward to spring. The sun, as always, goes on turning hydrogen into helium. We lean up toward the sun, tired of winter, greedy for our share of the missing 4 million tons.

APRIL 2

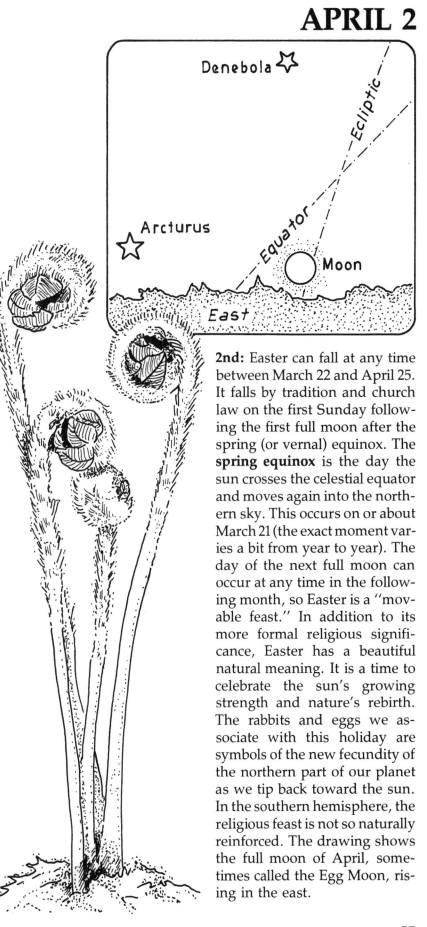

2nd: Easter can fall at any time between March 22 and April 25. It falls by tradition and church law on the first Sunday following the first full moon after the spring (or vernal) equinox. The **spring equinox** is the day the sun crosses the celestial equator and moves again into the northern sky. This occurs on or about March 21 (the exact moment varies a bit from year to year). The day of the next full moon can occur at any time in the following month, so Easter is a "movable feast." In addition to its more formal religious significance, Easter has a beautiful natural meaning. It is a time to celebrate the sun's growing strength and nature's rebirth. The rabbits and eggs we associate with this holiday are symbols of the new fecundity of the northern part of our planet as we tip back toward the sun. In the southern hemisphere, the religious feast is not so naturally reinforced. The drawing shows the full moon of April, sometimes called the Egg Moon, rising in the east.

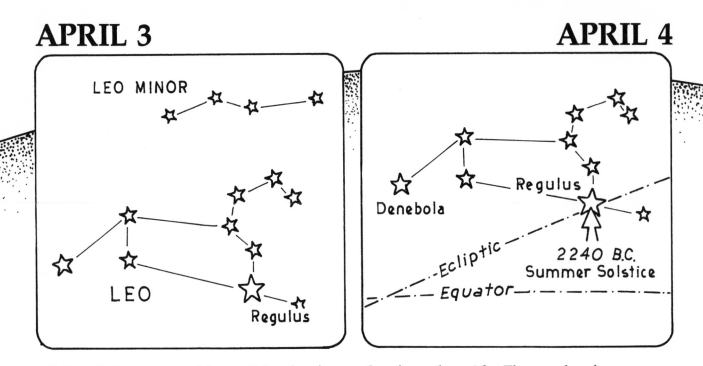

3rd: Leo the Lion is one of those groups of stars that is easy to associate with the mythological figure it is supposed to represent. Of the first-magnitude stars, Regulus stands alone in this part of the sky. No other nearby star rivals its brightness, so you should have no trouble finding it (beware of planets!). Above Regulus is the semicircle of stars that represents the majestic head of the lion. To the east is the easily identified triangle of stars that form the lion's hindquarters and tail. I like to imagine the lion in a reclining position, astride the ecliptic, guardian of its own broad expanse of spring sky. The star Regulus and the associated constellation have since ancient times been associated with the concept of "kingship." In this part of the sky, with few bright constellations, Leo is most certainly "king of the beasts."

4th: Thousands of years ago, the sun was in Leo when it was at its northernmost point in the sky. The **summer solstice** was very near Regulus in 2240 B.C. Even then this group of stars was called the Lion. The lion was known as an animal that liked the warm sun and the heat of the desert. It did not require a great leap of the imagination to associate the scorching summer sun with the lordly beast of the desert.

"The Sickle"

LEO

Eta Leonis

Regulus

LEO

Algeiba

Regulus

Eta Leonis

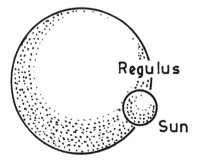

5th: An easy figure to find in the Lion is the "sickle." It is even easier to look for a "backward question mark." These are the stars that form the front part of the Lion. All of the stars in the "sickle" are at different distances from the earth. Regulus, for example, is about 85 light years away. Eta Leonis (*AY-ta lee-OH-nis*) is almost 25 times

Regulus

Sun

more distant. Regulus is a large star; you can see below its size compared to our sun. But Eta is a star of staggering proportions. Eta is a hot white supergiant star, and, because of both size and temperature, tremendously bright. On these two pages I have shown the size of Eta on the same scale as Regulus and the sun. If it were at the distance of Regulus, Eta would be the brightest star in our sky. If it were at the distance of Sirius (9 light years) it would be 50 times brighter than Venus and visible even in the daytime! Actually our starry nights are rather drab compared to what they might be if there were a few Eta-like stars in our neighborhood; or if the sun were a member of a multiple-star system; or, perhaps best of all, if our sun were embedded in a Beehive-like cluster of hot, young stars! But it is foolish to wish for what might have been—the sky we do have is a thing of astonishing beauty.

6th: Regulus is a **main sequence** star, both hotter and brighter than the sun. Certainly, even to the eye it has a bluish cast. With a telescope you can see that Regulus has a small companion, an orange star half as bright as the sun. The companion is itself a close double star, and circles Regulus at a distance 100 times greater than that at which Pluto orbits the sun. If you lived on a planet 93 million miles from Regulus, the orange binary companion would appear only as bright as Venus in our own sky. Algeiba (*al-GEE-ba*), which means "lion's mane," is also a binary star, available to the owner of a small telescope. One star book calls this binary pair "the dazzling golden duo."

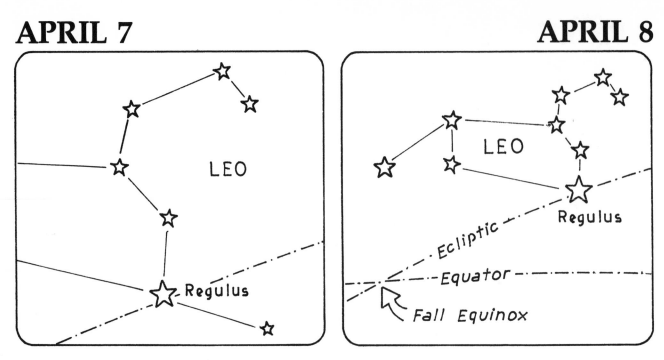

7th: It was Copernicus, the 16th-century astronomer who taught that the sun was not the center of the universe, who gave the brightest star in Leo, Regulus, its modern name. The name means "little king," and reflects the ancient belief that this was one of the royal stars which ruled the heavens. Many writers believe that the sphinx of Egypt, a figure with a human face and the body of a lion, was meant to associate the royal power of the Egyptian kings with the impressive celestial symbolism of Leo the Lion. This symbolism may be related to the fact that Regulus once marked the place of the **summer solstice.** It was by studying the thousand-year-old Babylonian observations of Regulus and Spica, two stars near the ecliptic, that Hipparchus of Alexandria discovered (in the 2nd century B.C.) that the positions of the equinoxes and solstices had changed.

8th: Leo is one of the twelve constellations of the **zodiac,** lying between Cancer and Virgo. Regulus is less than a degree off the **ecliptic.** When the sun comes this way on August 23 it just misses covering that star. The presence of Spica and Regulus on this part of the celestial sphere makes it easy for the stargazer to trace the ecliptic across the starry night. You will often see the moon pass very close to Regulus, and occasionally the moon will **occult** the star. If you see a bright "star" in Leo in addition to Regulus, it will be a planet. In 1959, Venus passed directly over the star, a very rare event.

Sphinx

60

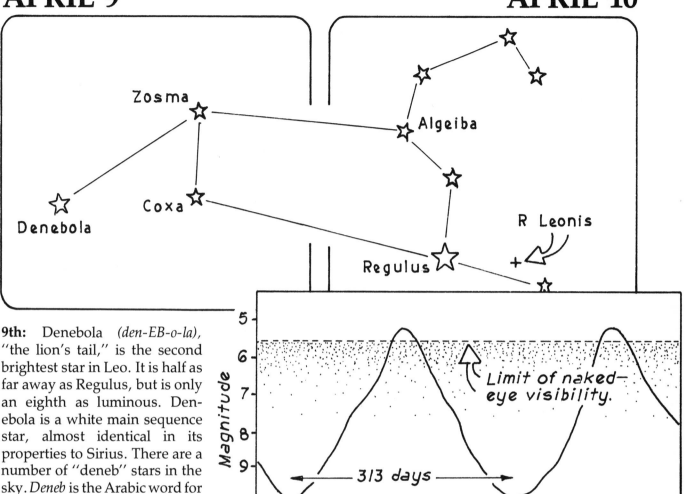

Zosma

Denebola

Coxa

Algeiba

R Leonis

Regulus

Magnitude

5

6

7

8

9

10

Limit of naked-eye visibility.

313 days

June 25, 1981

May 4, 1982

9th: Denebola *(den-EB-o-la)*, "the lion's tail," is the second brightest star in Leo. It is half as far away as Regulus, but is only an eighth as luminous. Denebola is a white main sequence star, almost identical in its properties to Sirius. There are a number of "deneb" stars in the sky. *Deneb* is the Arabic word for "tail." There is a "tail of the goat" (Deneb Algiedi in Capricorn), a "tail of Cetus the Whale" (Deneb Kaitos), a "tail" of the Dolphin, and the most famous deneb of all, the brilliant first-magnitude star at the tail of the summer constellation Cygnus the Swan. These names emphasize the close connection of many star names with the ancient mythological figures of the constellations. The two stars which make a triangle with Denebola have the curious names Zosma and Coxa. The former means "girdle" or "loincloth," and the latter means "hip." You can make of these star names what you wish, but it is difficult to imagine the lion in a pair of pants.

10th: Here is a rare and difficult object to look for with the naked eye. I will confess at once that I have never seen it. The star R Leonis, close by the paw of the lion, is a variable star that just surfaces the limit of naked-eye visibility when it is at its brightest. Usually it is too faint to be seen without binoculars or a small telescope. But about once a year it grows in brightness to the point where it should be barely visible to the eye under fine conditions. (Of course, the "limit of naked-eye visibility" is a somewhat inexact concept.) R Leonis is a variable star similar to the more famous Mira the Wonderful in Cetus (see Nov. 27). Thousands of such stars have been recognized. It seems that the light variations in these red giants are the result of a periodic expansion and contraction in the size of the stars. This "heavy breathing" is apparently related to instabilities in the balance between gravity and fusion as the stars use up their last energy resources.

APRIL 11

APRIL 12

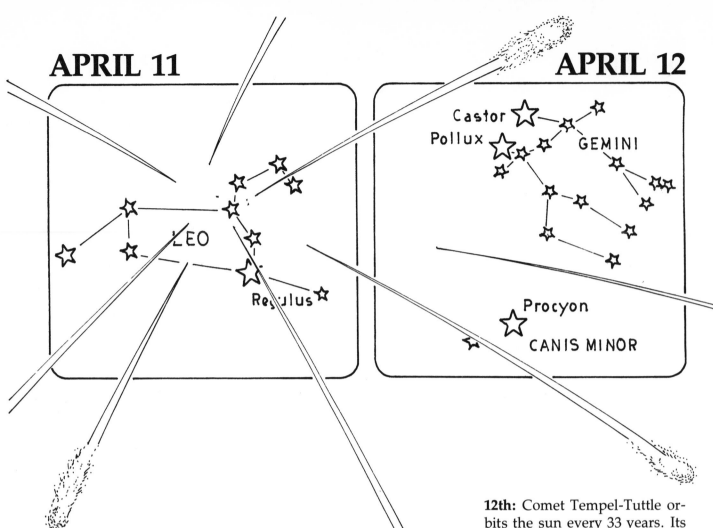

11th: Just to the northwest of the star Algeiba in Leo is the **radiant** of the meteor shower that takes its name from that constellation, the Leonids. **Meteors,** often called "shooting stars" or "falling stars," are typically sand-sized solid particles that plunge into the earth's atmosphere and are vaporized by the heat of friction. We see the luminous vapor as a brief flash of light across the dark night sky. You can see a meteor on almost any night if you look at the sky long enough, but at certain times of the year we are treated to grand showers of "falling stars." The Leonid shower peaks in the middle of November. On any clear night at that time, especially if you are up before dawn, you are likely to see a dozen or so meteors per hour. The Leonid shower results from a swarm of particles in orbit about the sun associated with the orbit of the comet Tempel-Tuttle. As the earth, in its own journey around the sun, passes through this stream of cometary litter, a shower results as particles collide with the earth's atmosphere (see Aug. 12). Meteor showers take the name of the constellation from which they appear to radiate. That area of the sky is called the shower's **radiant.**

12th: Comet Tempel-Tuttle orbits the sun every 33 years. Its long journey takes it out beyond the orbit of Uranus and back. Traveling behind the comet is a huge train or swarm of particles. When the comet swings in close to the earth, and if the earth passes through its train, spectacular meteor showers can result. These grander-than-usual showers can occur every 33 years. There was a wonderful shower of "shooting stars" following the comet's last visit in 1966. In some places over a hundred meteors per second were observed. There were also dazzling "fireworks" in 1799, 1833, and 1866. The next Leonid display is expected in the early morning hours of November 18, 1999. One wonders what primitive man thought of these showers of "stars." Did he think the sky was falling?

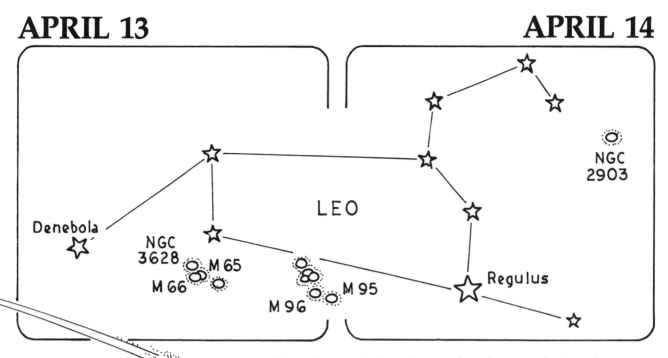

13th: Invisible to the naked eye, but spectacular on photographic plates produced at the large observatories, are the several clusters of galaxies in the hind quarters of Leo the Lion. Messier saw several of these galaxies as blurs of light and listed them in his catalogue. Of course, he could never have dreamed that each of the nebulous patches of light were whole "universes" of stars as numerous as those in the Milky Way. The clusters contain a number of fine spiral galaxies, and seeing them in a photograph we can imagine what our own Milky Way Galaxy would look like if viewed from some outside vantage point. In fact, we know less about the overall structure of our own galaxy than we do of the galaxy illustrated at right, for the same reason that a person in a huge crowd of people has difficulty knowing the shape and dimensions of the crowd.

14th: Near the head of the lion, there is another particularly fine spiral, NGC 2903 (NGC for *New General Catalogue*). How many stars are there in the galaxies of Leo? Each of the beautiful spiral galaxies contains hundreds of billions of stars. If we count only the eleven galaxies shown on the map, there are almost certainly over 1,000,000,000,000 stars! The eleven galaxies on the map represent only a tiny fraction of the many billions of galaxies that exist in the universe.

NGC 2903

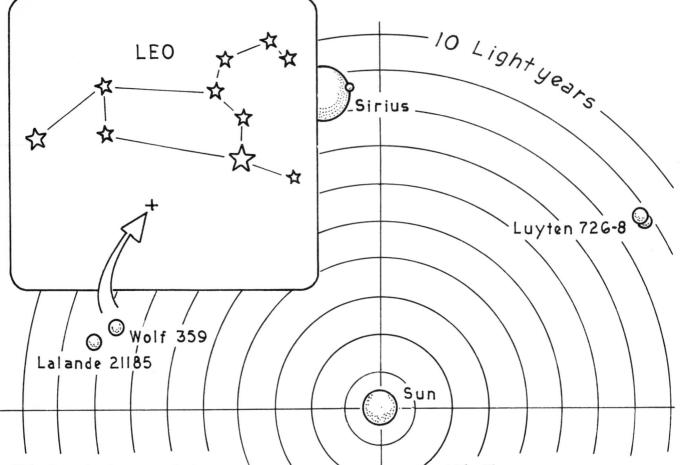

15th: Low in the constellation Leo, beneath the belly of the Lion, is our third nearest stellar neighbor in space, the star Wolf 359. The star is named for its discoverer, Max Wolf, and is visible only through a good-sized telescope. It is 7.7 light years away, not quite twice as distant as our nearest neighbor, Alpha Centauri. Wolf 359 is a tiny red star, at the very bottom of the **main sequence.** It is possibly not much larger than Jupiter, although certainly more massive. These little **red dwarf** stars are the most populous kind of star in the Galaxy. For every yellow star like our sun in the middle of the main sequence, there may be ten times as many red dwarfs.

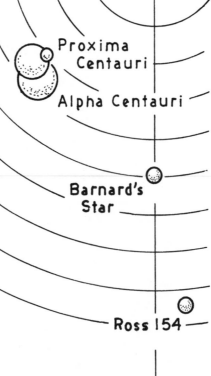

16th: There are seven star systems closer to the sun than 10 light years. These are the stars in our neighborhood of the Milky Way Galaxy—"on our block," so to speak. Of these stars, Sirius is the biggest and the brightest. One of our neighbors, Alpha Centauri, is a triple-star system. The smallest and closest of these three stars is called Proxima Centauri. Sirius has a **white dwarf** companion, and Luyten 726-8 is a pair of M-type stars. Barnard's Star and Lalande 21185 seem to have sizable unseen companions, probably large planets. It is not yet possible to physically see planets around nearby stars. Large space telescopes, in orbit around the earth, may make this possible.

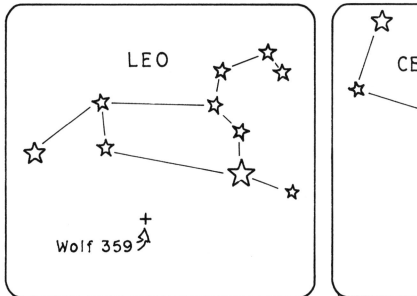

LEO

Wolf 359

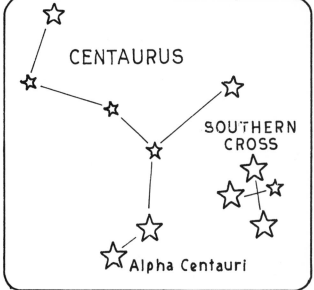

CENTAURUS

SOUTHERN CROSS

Alpha Centauri

17th: Of the eleven stars closer than 10 light years, the sun is one of four giants. But, as we have seen, compared to some more distant stars, like Rigel or Betelgeuse, even Sirius is a midget. The supergiants are relatively rare in the Galaxy, even though they tend to dominate our starry nights. If our neighborhood is typical, and there is no reason to believe that it isn't, the little **red dwarfs** are far and away the most populous kind of star. But they radiate so little light, we can see them only if they are nearby.

18th: Here is a piece of the sky unfamiliar to most stargazers in the northern hemisphere. You would have to go as far south as Key West or Honolulu to see it (see Sept. 29). In one small area are five of the heaven's thirty brightest stars. The brightest of these is our nearest neighbor in space, Alpha Centauri (*AL-fa sin-TAH-ree*). Alpha Centauri has three components. Alpha Centauri A and B are not unlike our sun. They are about as far

from each other as the sun and Uranus. The distance varies from 11 to 35 times the distance of the earth from the sun. Alpha Centauri C, called Proxima Centauri, is a miniscule dwarf star, far removed from the larger pair.

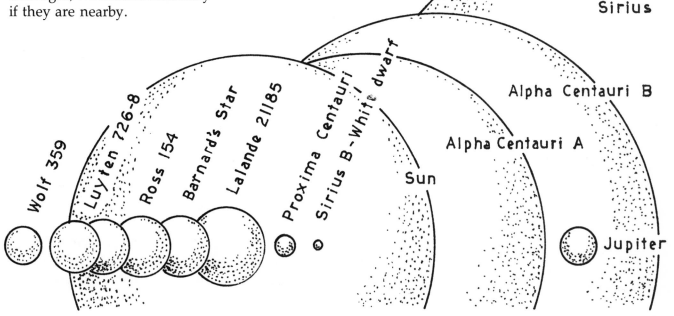

Wolf 359 Luyten 726-8 Ross 154 Barnard's Star Lalande 21185 Proxima Centauri Sirius B-White dwarf Sun Alpha Centauri A Alpha Centauri B Sirius Jupiter

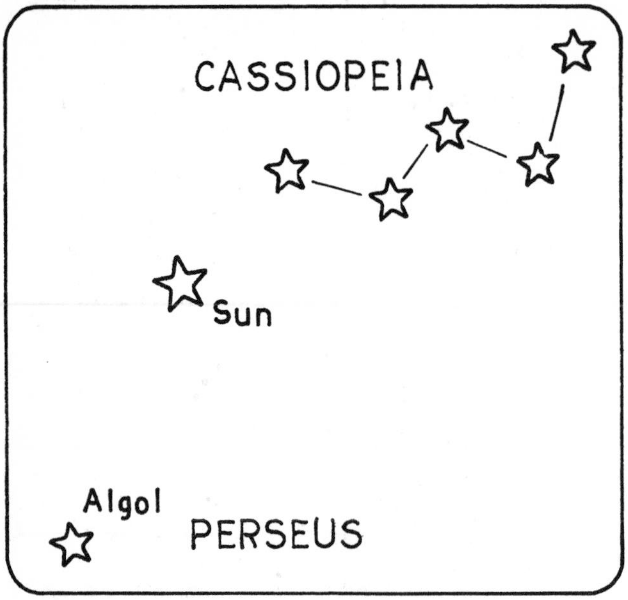

CASSIOPEIA

Sun

Algol

PERSEUS

CENTAURUS

SOUTHERN CROSS

Alpha Centauri

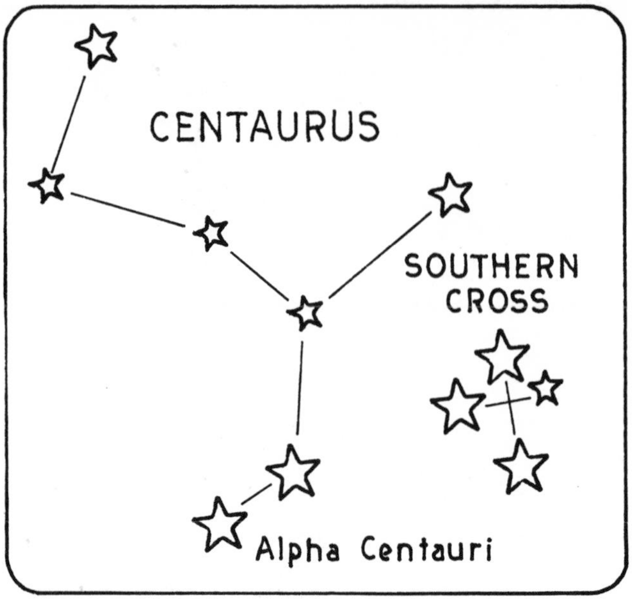

19th: If we are not alone in the universe, then where should we begin searching for other creatures? Probably near stars similar to our own sun. Alpha Centauri A is just such a star, an almost perfect twin of the sun. Such stars are bright enough to warm an appreciable volume of their surroundings, and have long enough lifetimes to allow for the evolution of higher forms of life. If there are citizens of Alpha Centauri A, orbiting that star on an earthlike planet, they would see our sun as one of the brightest stars in their sky, brighter even than the third member of their own star system, Proxima Centauri. More specifically, the sun would be a little less bright than Alpha Centauri appears to us. It would be located near the group of stars we call Cassiopeia.

20th: Of course, no star in the Alpha Centaurian's sky would rival the companion of their own "sun." Alpha Centauri B, a brilliant orange star, would have a diameter about 1/20th that of the nearer yellow "sun," and be far brighter in their sky than the moon is in our own. It would make the circuit of their heavens in 80 years, providing dramatically changing patterns of light. "Sunrise" on our imaginary earthlike planet would sometimes feature two "rising suns," something like what is pictured below.

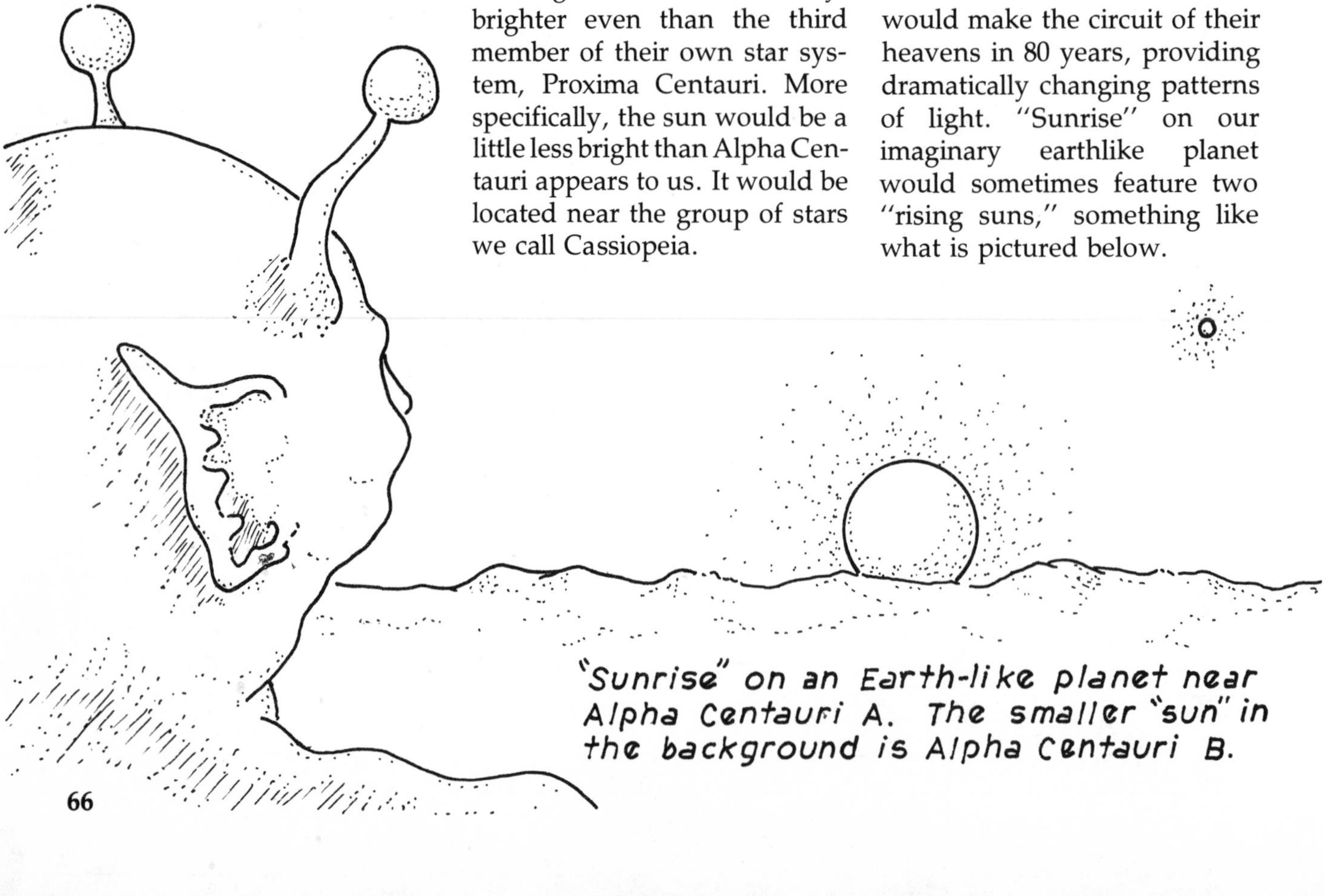

"Sunrise" on an Earth-like planet near Alpha Centauri A. The smaller "sun" in the background is Alpha Centauri B.

APRIL 21

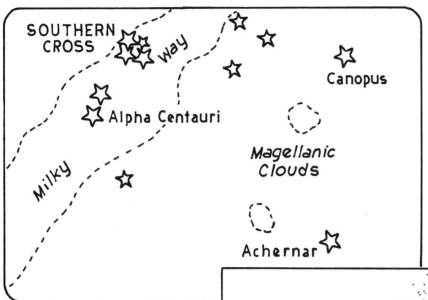

APRIL 22

22nd: In no other part of the sky are there so many first-magnitude stars so close together. Ten of the thirty brightest stars in the heavens are visible in the view below. Of these, only Sirius, high in the northwest, and its neighbor Adhara to the east, are generally familiar to mid-northern observers. And don't forget—in the southern hemisphere the season is fall; the water at your feet still holds summer's heat.

21st: While we are in southern skies and imagining the view from distant places, let us consider what the sky would look like this starry night from a beach in Australia. The scene pictured above and at right is toward the south (the south celestial pole is near the center of both drawings). It shows part of the celestial sphere that is inaccessible to the northern observer. No starry night anywhere on the planet can be quite as spectacular as this. The Milky Way rises out of the sea and soars toward the zenith. To the west we see the Large and Small Magellanic Clouds, small satellites of the Milky Way Galaxy. They are classed as irregular galaxies and lack the elegant symmetry of the spirals or the ellipticals. Each contains tens of billions of stars. The Magellanic Clouds were discovered by Portuguese sailors who rounded the tip of Africa in the 15th century, and were later named for the great circumnavigator who also sailed through southern seas.

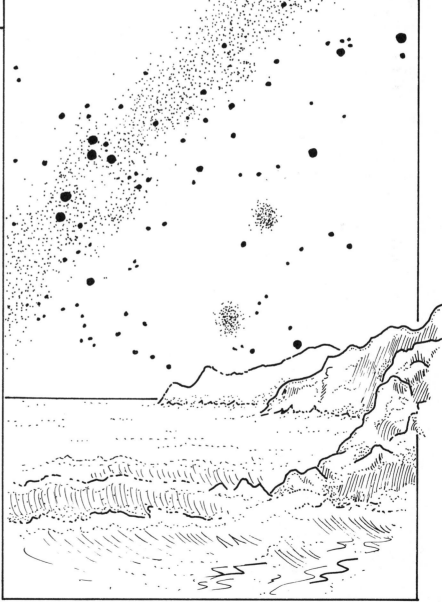

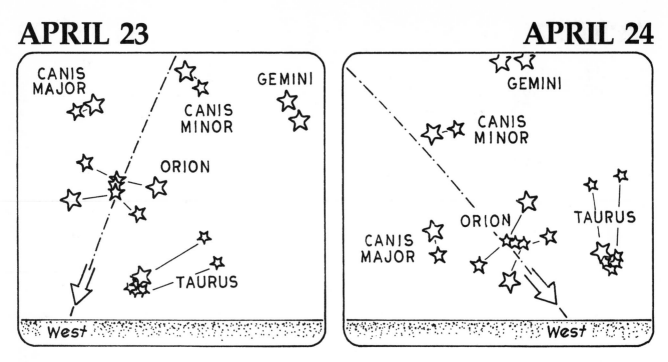

23rd: Let us have one more armchair "starry night" before we go back to northern skies. The map shows familiar northern constellations as they would appear from Australia this evening after sunset. The sketch below shows the same scene. The view is toward the west and includes a few of the stars shown on the preceding page.

The view can be disorienting for northern stargazers. For northern observers, Orion sets on his feet; our imaginary Australian sees him dive head first into the sea following the bull. The bright star above the ship's mast is Canopus, the second brightest star in the heavens. At the ship's stern is Achernar, tenth on a list of the brightest stars.

24th: Back to the north. This is what the western horizon looks like tonight as the sky darkens, for an observer at latitude 40°N. It is a topsy-turvy version of the Australian scene below. We see Orion and Taurus go down together, still battling it out until only the tips of the bull's horns remain above the horizon.

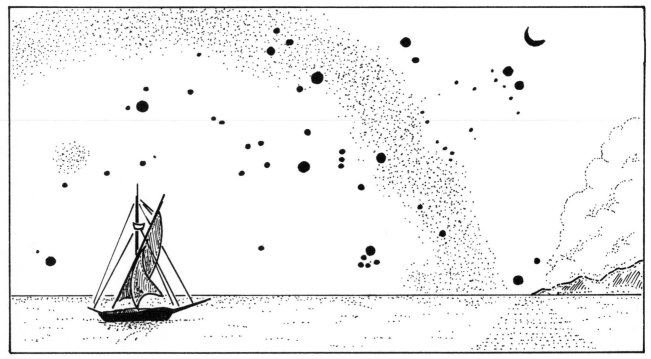

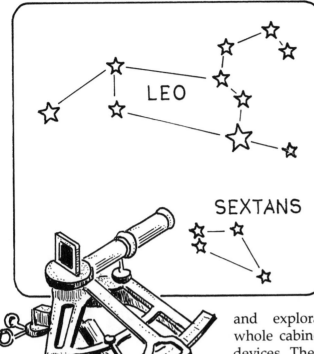

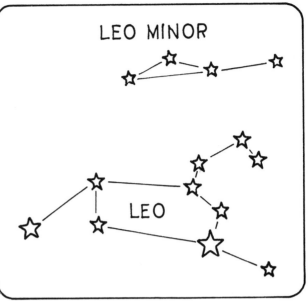

25th: Here is a little constellation with no notable stars. In older times, the brightest star in a constellation was called its *lucida*. The lucida of Sextans is a star of only the 4th magnitude. The constellation was created and named by the 17th-century astronomer Hevelius, who was interested among other things in the nature of comets. The tiny stars in this area of the sky had previously gone without designation, and ("with more zeal than taste," as one writer said) Hevelius named them for a favorite astronomical instrument. The constellations of the southern skies, mostly invented after the great age of discovery and exploration, contain a whole cabinet of technological devices. There is a sextant and an octant, a telescope and a microscope, a clock and a ship's compass, a pump and a furnace. Most of these received their names from Nicolas Louis de Lacaille in the 18th century. It is important to remember that before Europeans observed and named the stars of the southern skies, the inhabitants of those climes had their own rich mythologies of celestial lore.

26th: Here is another of the constellations named by Hevelius, who was intent on claiming the faint unnamed stars between the traditional constellations. The brightest star of Leo Minor is also of the 4th magnitude, so you will need an excellent evening to recognize the little lion. Hevelius was of a less technological bent than Lacaille; most of his invented constellations are small animals. He added a little lion, a little fox, a lynx, a lizard, and a pair of dogs to the celestial zoo.

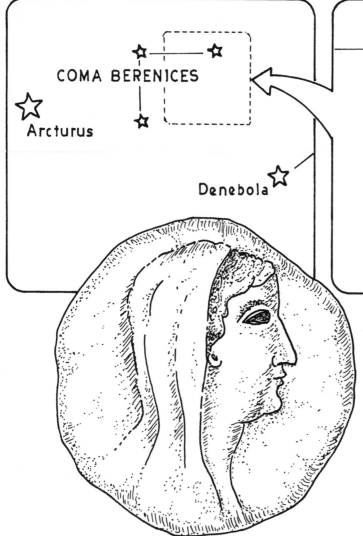

COMA BERENICES

Arcturus

Denebola

NGC 4565

COMA BERENICES

Coin of 3rd century B.C. showing likeness of Berenice.

27th: Leo is surrounded on four sides by inconspicuous constellations: to the west, Cancer; to the south, Sextans; to the north, Leo Minor; and to the east, the faint array of stars known as Coma Berenices, "Berenice's lock of hair." The constellation was officially catalogued by the great Tycho Brahe in 1602, but its origin is very ancient. Berenice is a historical person, the only nonmythological human honored in the name of a constellation. She was the daughter of the King of Cyrene, and became Queen of Egypt as the wife of Ptolemy III. She lived in the 3rd century B.C. According to legend, Berenice cut off her beautiful hair in thanksgiving to the goddess Venus for the safe return of her husband from war. Despite such a display of affection, her husband was not pleased when he returned to his shorn spouse. His anger was allayed only when the court mathematician, Conon, proclaimed that the gods had removed Berenice's tresses from the temple of Venus and placed them among the stars.

28th: This scattered group of faint stars was more familiar in times past, when skies were brighter than they are today. The stars of Berenice's locks are part of a true cluster, lying at a distance of about 250 light years. It is one of the nearest of the star clusters (the Pleiades are at a distance of 400 light years). There are about a half-dozen stars accessible to the naked eye, but the cluster is more impressive with binoculars. Coma Berenices harbors many fine galaxies. The modern telescope reveals a dazzling array of impressive spirals, each a system of stars as large as the Milky Way Galaxy. NGC 4565 is a particularly interesting specimen, a great spiral seen exactly edge-on, as thin as a dinner plate except for the central bulge of the nucleus. Another of the beautiful Coma Berenices spirals is shown on July 13.

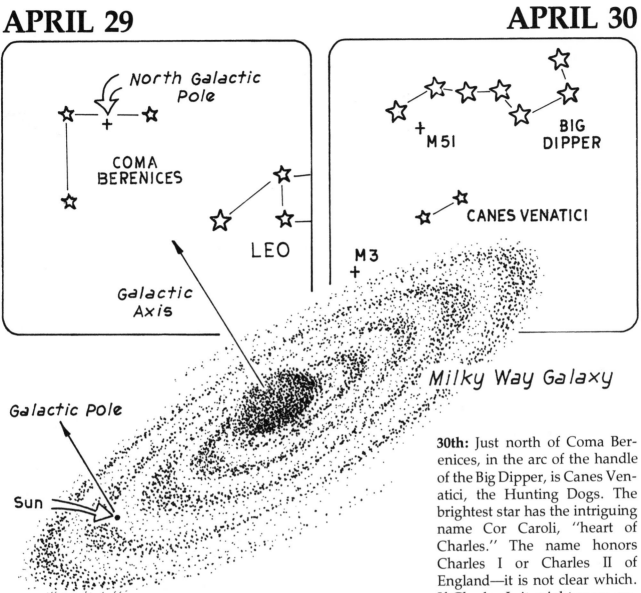

North Galactic Pole

COMA BERENICES

LEO

Galactic Axis

Galactic Pole

Sun

M 51   BIG DIPPER

CANES VENATICI

M 3

Milky Way Galaxy

29th: Although the constellation Coma Berenices lacks bright stars, it is rich in telescopic objects. Globular star clusters and galaxies abound. The constellation is also the location of the north **galactic pole,** the place where a line parallel to the rotation axis of the Milky Way Galaxy intersects the celestial sphere. This is the part of our sky furthest from the Milky Way. When we look toward the stars of Coma Berenices, we are looking directly up and out of the flat disk of the Galaxy. This fact accounts for the poverty of bright stars in this part of the sky. It also permits a relatively unobstructed view of the many fine extragalactic objects to be found in Coma Berenices. Ancient astronomers once imagined the Coma Berenices cluster of stars as the tuft of hair at the end of Leo the Lion's tail. The universe of those high and far-off times was compact and human-centered. Telescopic photographs of deep-sky objects in Coma Berenices have made us citizens of the vast universe of the galaxies.

30th: Just north of Coma Berenices, in the arc of the handle of the Big Dipper, is Canes Venatici, the Hunting Dogs. The brightest star has the intriguing name Cor Caroli, "heart of Charles." The name honors Charles I or Charles II of England—it is not clear which. If Charles I, it might more appropriately be called *Caput Caroli* ("head of Charles"). There are two magnificent telescopic objects in the constellation. M 3 is one of the most beautiful of the **globular clusters** (see July 5–7). Best of all is spectacular M 51, the *Whirlpool Galaxy,* a face-on spiral. This was the first nebula recognized to have a spiral structure, by Lord Rosse in 1845. Rosse used a huge six-foot reflecting telescope which he had constructed on his estate in Ireland. Not until 1924 was the nature of the spiral nebulae understood.

THE EDGE OF NIGHT
April 30

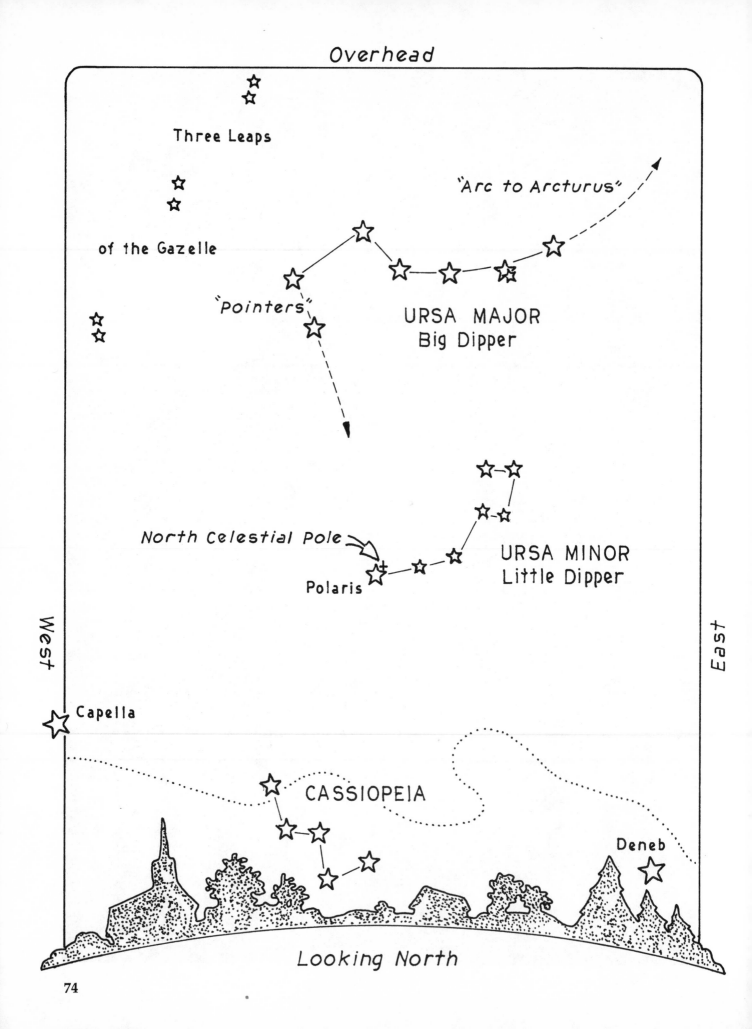

Overhead

Three Leaps

of the Gazelle

"Arc to Arcturus"

"Pointers"

URSA MAJOR
Big Dipper

URSA MINOR
Little Dipper

North Celestial Pole

Polaris

West

East

Capella

CASSIOPEIA

Deneb

Looking North

MAY 1

1st: Deep beneath the winter leaf litter that covers the forest floor the earth responds to the increasingly direct radiation of our yellow star. The first wildflowers of the new season push up green leaves through the brown debris. Each leaf is a complex chemical factory designed to convert carbon dioxide, water, and starlight into the fuels of life. Energy that had its source deep in the sun's core now finds expression in the delicate pink folds of the lady-slipper orchid and the tiny white flower heads of the wild lily-of-the-valley. Days in May are marvelous for walks in the woods, for watching the sun

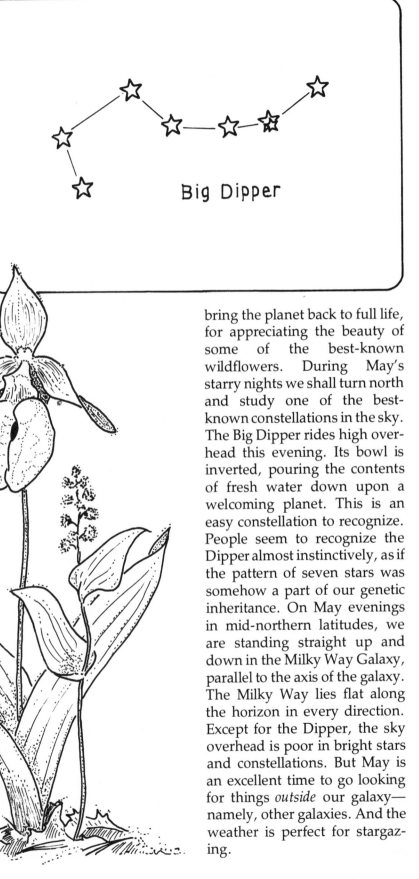

Big Dipper

Ladyslipper and
Wild Lily-of-the-Valley

bring the planet back to full life, for appreciating the beauty of some of the best-known wildflowers. During May's starry nights we shall turn north and study one of the best-known constellations in the sky. The Big Dipper rides high overhead this evening. Its bowl is inverted, pouring the contents of fresh water down upon a welcoming planet. This is an easy constellation to recognize. People seem to recognize the Dipper almost instinctively, as if the pattern of seven stars was somehow a part of our genetic inheritance. On May evenings in mid-northern latitudes, we are standing straight up and down in the Milky Way Galaxy, parallel to the axis of the galaxy. The Milky Way lies flat along the horizon in every direction. Except for the Dipper, the sky overhead is poor in bright stars and constellations. But May is an excellent time to go looking for things *outside* our galaxy—namely, other galaxies. And the weather is perfect for stargazing.

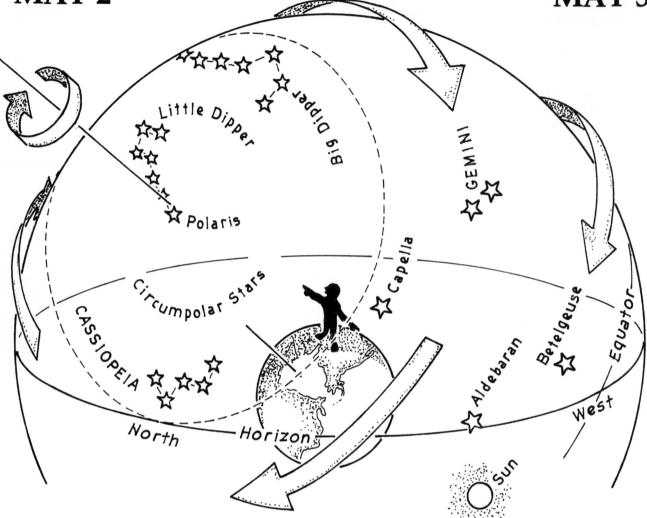

2nd: The earth turns west to east on its axis under the stars. This makes the sky seem to turn east to west over the observer. Accordingly, the stars rise in the east and set in the west. The drawing shows the sky this evening as viewed from *outside* the imaginary **celestial sphere.** Our North American observer has his arms outstretched parallel to the earth's axis, his right hand pointing to Polaris. The sun has sunk below the horizon and the sky has darkened. Aldebaran is setting in the west. Betelgeuse will soon follow. These were the stars that were high overhead on January evenings. Meanwhile, the first-magnitude stars of summer, Vega and Deneb, are rising in the east. All of the stars rise and set as the earth turns, except those stars closer to the pole than the observer's northern horizon. These **circumpolar stars** are always in the sky. They are there even in the daytime, though lost in the sun's brilliant light. Cassiopeia, for example, will not set for our observer. As the night passes, the constellation will skim the northern horizon and be high in the north-eastern sky at sunrise. During the daylight hours it will pass overhead, and reappear low in the northwest as the sun sets and the stars "come out."

3rd: If you go out this evening you will find the Big Dipper high overhead. Line up several of its stars with the edge of your house or some other stationary reference. Now go inside and watch TV for a few hours. When you look for the Dipper again you will see that it has swung to the west, in a great arc about the pole, 15° for every hour you were away. By dawn it will be low in the northwest, just above the horizon. The stars of the Dipper seem such delicate points of light. It is not hard to understand why the ancients believed it was the sky that turned rather than the earth.

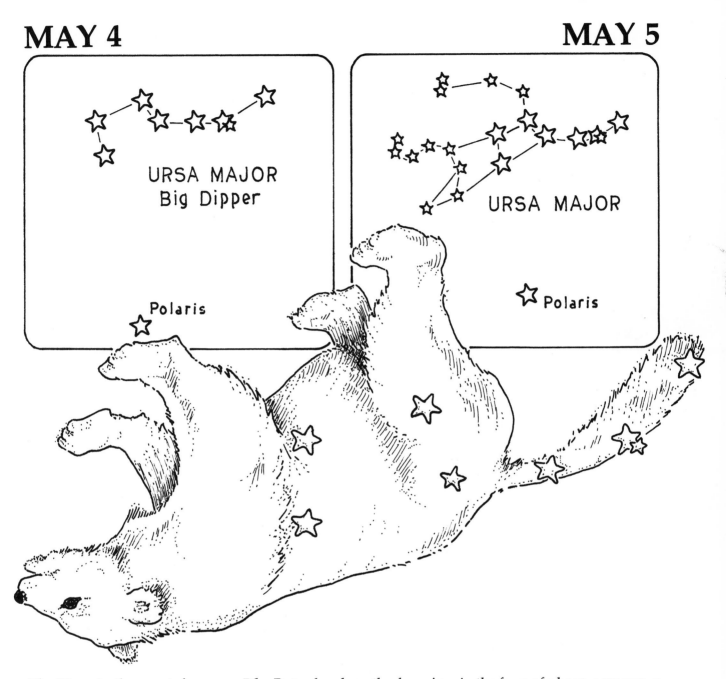

URSA MAJOR
Big Dipper

Polaris

URSA MAJOR

Polaris

4th: Here is the most famous and familiar of all constellations, the Big Dipper, or Ursa Major, the Great Bear. No other group of stars, save possibly Orion, is easier to recognize than this constellation. In many cultures all around the globe these stars have been associated with the figure of a bear. The universality of this designation suggests that this may be one of the most ancient of all named constellations.

5th: But why does the bear in the sky have such an absurdly long tail? The picture I have drawn here is purely imaginary. No bear alive on the planet today has a tail so long. Did the bears hunted by our Stone Age ancestors have tails as long as the creature they put in the sky? If nothing else, the long-tailed Great Bear confirms the resourcefulness of the human imagination. Indeed, these seven bright stars have been imagined

in the form of a bear, a wagon, a plough, and most recently a dipper. It is as a dipper pouring fresh water down upon an awakening earth that we see the constellation this starry night. Once seen, its form is never forgotten. On the map above, I have included a few of the less conspicuous stars of the constellation and connected them in such a way as to suggest a little more vividly the figure of a bear.

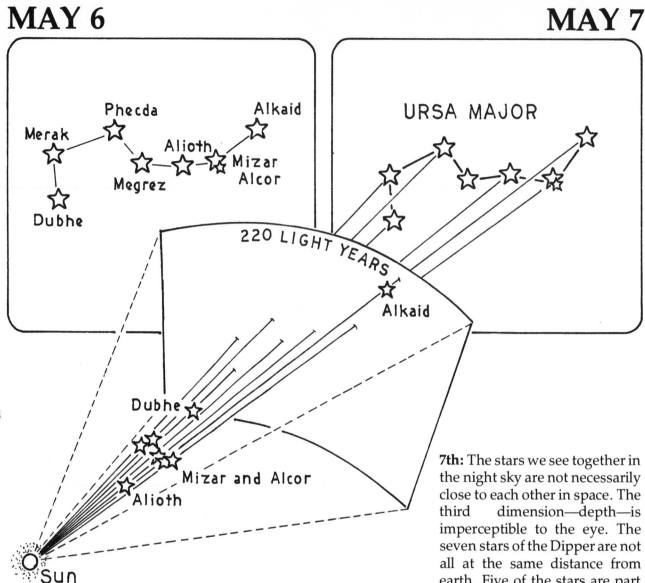

Merak · Phecda · Alkaid · Alioth · Megrez · Mizar Alcor · Dubhe

URSA MAJOR

220 LIGHT YEARS
Alkaid

Dubhe
Mizar and Alcor
Alioth
Sun

6th: Zeus, the greatest of the gods, fell in love with the beautiful mortal Callisto. Callisto was a huntress who roamed the mountains of Acadia in search of game. Hera, the wife of Zeus, was intensely jealous of her husband's new obsession and changed the unfortunate Callisto into a bear. As a bear, Callisto cowered in the forest terrified of both man and beast. One day her son, Arcas, came upon the bear. Happy Callisto stood on her hind legs to welcome him! Thinking himself attacked, Arcas readied his bow.

But Zeus, looking down from high Olympus, saw the terrible event unfolding, and with instantaneous divine magic, changed Arcas into a little bear. Grabbing both bears by the tails (and stretching them?), Zeus hurled them into the safety of the sky. But Hera, as usual, had the last laugh. She moved the bears into that part of the sky near the celestial pole, where they would never set and therefore never rest. And there they are today, arched poignantly toward each other, eternal victims of Zeus' wandering eye.

7th: The stars we see together in the night sky are not necessarily close to each other in space. The third dimension—depth—is imperceptible to the eye. The seven stars of the Dipper are not all at the same distance from earth. Five of the stars are part of a true cluster, at a distance of about 80 light years. The nearest of the stars is Alioth (ALLEY-oth) at 70 light years. Orange Dubhe (DU-bee) lies just beyond the cluster at 105 light years. The blue-white giant Alkaid (al-KAID) is far beyond all the others, double the distance of Dubhe, at 210 light years. In spite of its great distance, Alkaid shines as brightly as its six brothers. All of the Dipper stars lie on or close to the **main sequence** (see Mar. 15), except Dubhe. Dubhe is an orange giant, and appears orange to the eye.

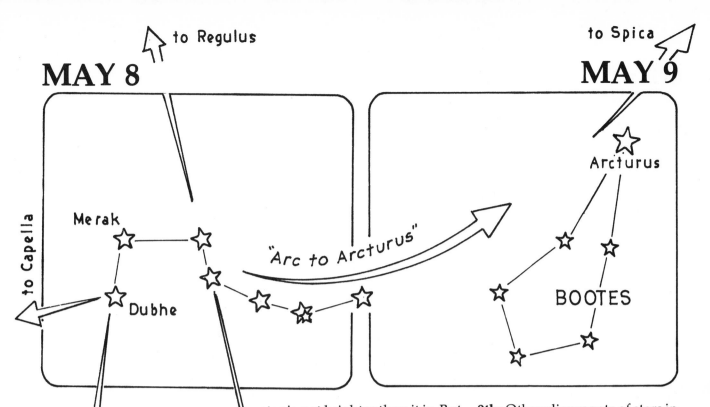

to Capella

Merak

Dubhe

"Arc to Arcturus"

Arcturus

BOOTES

"Point to Polaris"

Polaris

Deneb

8th: The Big Dipper is a terrific guidepost to the starry night. Most important, the Dipper can lead you straight to the North Star, Polaris. The trick is familiar to every Boy or Girl Scout, presumably to be used on that mythical occasion when we are lost in the woods at night. Merack (*ME-rack*) and Dubhe (*DU-bee*), the stars at the front of the bowl of the Dipper, are called the *Pointers.* Follow them out of the bowl and they will lead you *almost* directly to Polaris. Polaris will be about five times further away from Dubhe than the distance between the *Pointers.* Most people seem to be surprised to find that the north star is not brighter than it is. But we are lucky to have a pole star at all (see May 22–23), and Polaris is certainly the brightest star in its part of the sky. If you stand facing Polaris, your right arm points east and your left arm west. South is at your back. This evening the *Pointers* are twelve o'clock high on the great "clock" in the northern sky. Wait two hours and take another look. The *Pointers* will now be at eleven. The "clock" turns counterclockwise! The name Dubhe, by the way, is derived from the Arabic word for "bear." Merack means the "loin of the bear," and helps us visualize what it was the ancients saw in the sky.

9th: Other alignments of stars in the Big Dipper will lead you to Capella, to Regulus, and to Deneb in the summer swan. The most familiar of these guidelines takes you out along the handle of the Dipper. Continue along the arc of the handle and you will arrive at brilliant Arcturus, the fourth brightest star in the sky. Keep going along the same arc, maintaining the same curvature, and you will arrive at Spica (*SPY-ka*) in the constellation Virgo. Spica is a first-magnitude star almost exactly on the **ecliptic.** The rule to remember is this: "Make an *arc* to *Arc*turus, keep going and you will *spy Spi*ca." These guidelines can be especially useful on those nights when only the brightest stars are visible in the sky. Without the patterns of the familiar constellations as a help, it is sometimes difficult to recognize the bright stars. The Dipper guidepost is useful at any time of the night and at any time of the year.

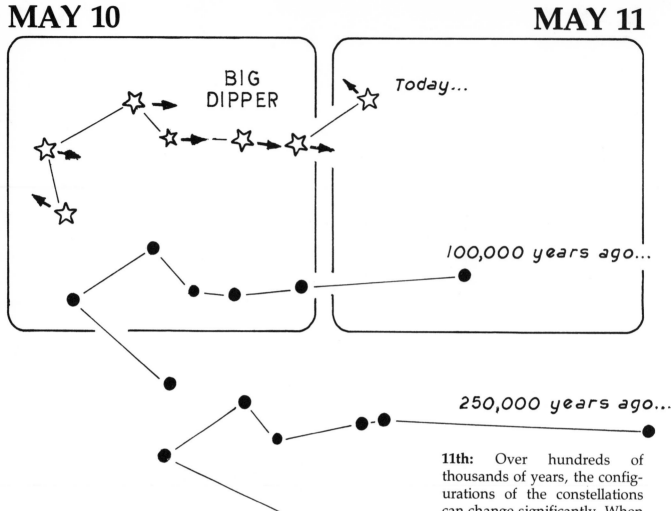

BIG
DIPPER

Today...

100,000 years ago...

250,000 years ago...

10th: The Dipper guidepost is "permanently" useful because stars maintain their positions *relative to each other* even as the earth moves. But not exactly! The stars do have individual motions within the Galaxy which cause their positions on the celestial sphere to slowly change (see Feb. 17). These changes are called **proper motions.** By photographing the stars over a long interval of time, astronomers can measure proper motion. For nearby stars the relative changes in position will appear the greatest. But even for nearby stars the changes are slight, typically a fraction of a second of arc per year. The stars of the Big Dipper require tens of thousands of years to change positions by even 1° on the celestial sphere. The five middle stars of the Dipper have similar proper motions across the sky, further evidence that these stars are part of a true cluster. Dubhe and Alkaid "race" off in the opposite direction. These proper motions will slowly distort our Big Dipper guidepost. Future generations will require new guidelines.

11th: Over hundreds of thousands of years, the configurations of the constellations can change significantly. When Swanscombe man looked up 250,000 years ago from his campsite in what is now England, what he saw in the sky may have looked more like a bear but certainly much less like a dipper (of course, he had no dippers anyway). When our Neanderthal ancestors, who painted the marvelous figures in the caves at Lascaux and Altamira, looked up to the starry night, the stars of the Dipper were beginning to assume their present pattern. In all of recorded history the changes have been but slight. The star maps in this book will be an adequate guide for your own great-great-grandchildren.

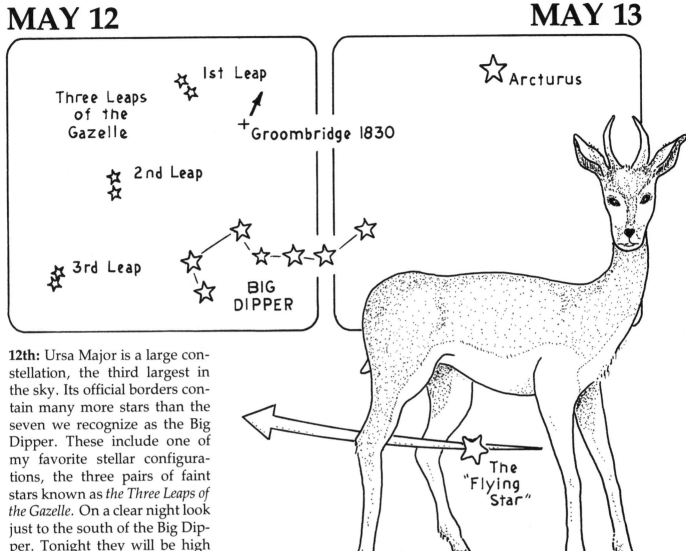

12th: Ursa Major is a large constellation, the third largest in the sky. Its official borders contain many more stars than the seven we recognize as the Big Dipper. These include one of my favorite stellar configurations, the three pairs of faint stars known as *the Three Leaps of the Gazelle.* On a clear night look just to the south of the Big Dipper. Tonight they will be high over your head. According to an ancient legend, the dark expanse of sky between the Great Bear and Leo the Lion is known as *the Pond.* The gazelle leaped into the pond to escape the lashing of the lion's tail. You will notice that the pair of stars called *the First Leap* is not far from Denebola. With a little imagination you will see the gazelle spring lightly away from danger, in three long leaps through the shallow pond. The legend does not tell us what happened to the gazelle on the other side of the pond. Let us hope that it did not encounter the bear!

13th: The second and third leaps of the gazelle are usually shown on the old star maps as the feet of the bear. But I find it best to associate the bear with the dipper stars only, and leave the three little pairs to their own delightful story. There is, by the way, a real "gazelle" in this part of the sky. Not far from the first leap is the telescopic star listed as Groombridge 1830. A more romantic name is *the Flying Star.* This star has a **proper motion** which is exceeded by that of only two other stars. It moves across the sky at a rate of 1° every 500 years (compared to tens of thousands of years for the stars of the Dipper), in a direction opposite to the leaping gazelle. Surprisingly, Groombridge 1830 does not turn out to be one of our nearest neighbors. It is 28 light years distant, and there are several hundred stars closer to us than that. This means that the flying star is flying not only through *our* sky, but has an unusually high *space velocity* compared to other stars.

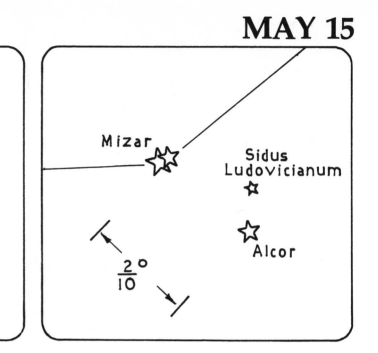

14th: At the bend of the Dipper's handle is the interesting pair of stars Mizar and Alcor. The pair is frequently called *the Horse and Rider*. American Indians named them *the Squaw and the Papoose*. Arabs considered Alcor a test of good eyesight, but you should have no trouble seeing the pair if the night is clear.

15th: There is another story about Alcor that I enjoy. The Pleiades in Taurus are often referred to as *the Seven Sisters* (see Jan. 23). Most people have difficulty seeing more than six stars in this beautiful cluster. What became of the *Lost Pleiad*? The story relates that the seventh sister was taken away by Mizar, one of the seven brothers of the Big Dipper, and there she remains, little Alcor, at his side. A small telescope will reveal Mizar to be a striking double star, the first binary star to be recognized as such. It is usually included on any telescope owner's grand tour of celestial sights. A spectroscope reveals that both components of Mizar are themselves double stars, too close to be separated by telescope. Near Mizar and Alcor is the curiously named Sidus Ludovicianum, Ludwig's star. The star was observed in 1723 by a German who thought he had discovered a new planet. He named it for his sovereign, Ludwig V, landgrave of Hesse-Darmstadt.

Horse and rider from 13th-century Islamic manuscript

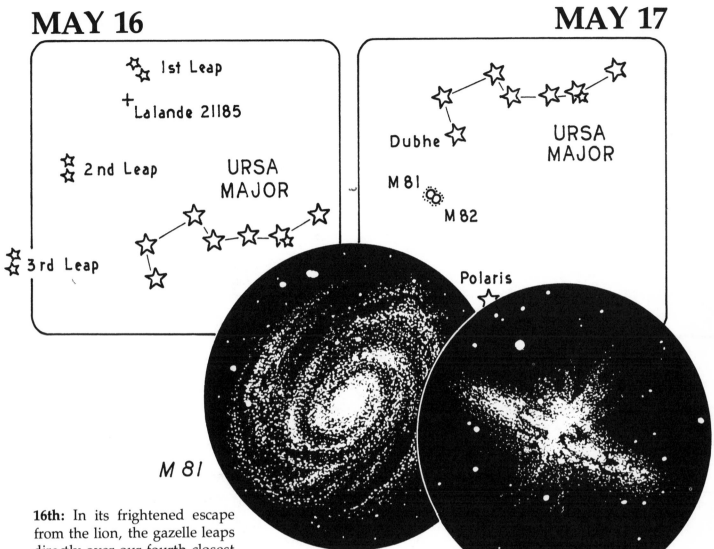

1st Leap

Lalande 21185

2nd Leap

URSA MAJOR

3rd Leap

Dubhe

URSA MAJOR

M 81

M 82

Polaris

M 81

M 82

16th: In its frightened escape from the lion, the gazelle leaps directly over our fourth closest stellar neighbor (see Apr. 15–16). Lalande 21185 is 8.3 light years away, just a little further than Wolf 359 in Leo and a little closer than Sirius. From slight variations in the proper motion of the star, it has been deduced that Lalande 21185 has a tiny unseen companion with a mass about 1/100th that of the sun. This puts the object on the borderline between tiny star and huge planet. It is unlikely, however, that we will find celestial neighbors on planets of Lalande 21185. The star is a **red dwarf,** incapable of heating a very large volume of space to temperatures comfortable for life as we know it.

17th: Ursa Major is home for a number of fine **galaxies.** Not far from the lip of the Dipper are two of the most interesting. They are easily available to the owner of a small telescope, although you will not see the details represented above. M 81 is a superb spiral galaxy, part of a group of galaxies about 7 million light years away. This is not far away as galaxies go (see Nov. 18). The galaxy is very similar to our own, a languid whirlpool containing tens of billions of suns. M 82 is a much more puzzling object. It lacks the elegant symmetry of M 81. The body of the galaxy is an amorphous ellipsoid. The central area is marked by wisps and filaments, as if some colossal explosion had occurred in the nucleus. Material streams outward from the nucleus at hundreds of miles per second. The galaxy is a strong source of radio energy and has a strong magnetic field. Whatever is going on in M 82, it is certainly one of the most violent events to have occurred in nearby regions of the universe.

URSA MAJOR

"Guard Stars"

URSA MINOR

Polaris

Kochab

URSA MINOR

Polaris

18th: People are usually disappointed when the Little Dipper is first pointed out. The stars are not very bright. Except for Polaris at the end of the handle and the two stars at the front of the bowl, they are difficult to see unless you have a very clear night. And the handle is bent curiously upward, not at all like a real dipper. But *Dipper* is a modern designation for this constellation. The official name is Ursa Minor, the Little Bear. Like Ursa Major, we must imagine this creature to have an absurdly long tail. The most interesting thing that can be said about the small bear is that the star at the tip of its tail is very near the north **celestial pole.** If you were standing at the north pole of the earth, this star would be almost exactly overhead.

19th: New stargazers often seem to expect the North Star to be one of the brightest in the sky. But this is not so. Polaris is a second-magnitude star, no brighter than the stars of the Big Dipper. An old story says that the Great Bear wanted Polaris for herself, since it matches her own stars in brightness. But the *Guard Stars* of the Little Bear, the two stars at the front of the Little Dipper, have protected Polaris and prevented the Great Bear from increasing her number to eight. It is pure chance that we have a pole star. In fact, Polaris is half a degree away from the pole. And, as we shall see, it has not always been so near.

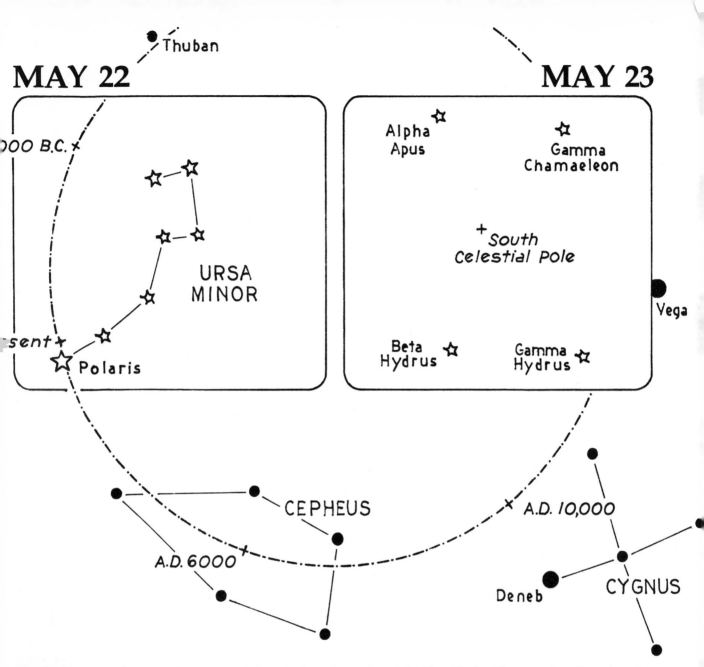

MAY 22

Thuban

000 B.C. ✕

☆ ☆

☆—☆

☆

URSA
MINOR

☆

☆

sent ✕

☆ Polaris

CEPHEUS

A.D. 6000

MAY 23

☆ Alpha
Apus

☆ Gamma
Chamaeleon

✛ South
Celestial Pole

Vega

Beta ☆
Hydrus

Gamma ☆
Hydrus

A.D. 10,000 ✕

Deneb

CYGNUS

22nd: Present-day navigators are fortunate to have a star to point the way north. It has not always been so. The earth's axis has a tilt of 23½° to the plane in which the earth orbits the sun —called the **plane of the ecliptic.** It is the tilt of the earth's axis that causes the seasons, as the planet's northern and southern hemispheres lean alternately toward the sun's face. Moreover, the axis of the earth wobbles, the way the axis of a top wobbles as the top spins. The effect is called **precession.** One circular wobble

of the earth-top requires 25,800 years. Presently the earth's axis points toward the distant star Polaris. Actually, the alignment of axis and star is improving. The **celestial pole** will make its closest approach to Polaris in the year A.D. 2102, at which time it will be a little less than ½° from that star. Five thousand years ago the earth's axis pointed toward Thuban in Draco, and that star was our "Polaris" (see Sept. 22). In A.D. 15,000 the pole will be near the first-magnitude star Vega.

23rd: Those who live in the southern hemisphere are not so fortunate in having a star to mark the southern pole. Although some of the most star-rich regions of the sky are in the southern hemisphere, few parts of the celestial globe are as poor in stars as that near the south celestial pole. The southern pole lies in the constellation Octans. The best we have for a "Polaris" of southern skies is the star Sigma Octans, 1° away from the pole. It is a star of magnitude 5.5, barely at the limits of naked-eye vision.

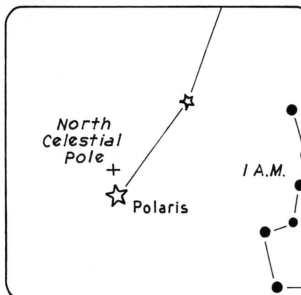

North Celestial Pole + Polaris

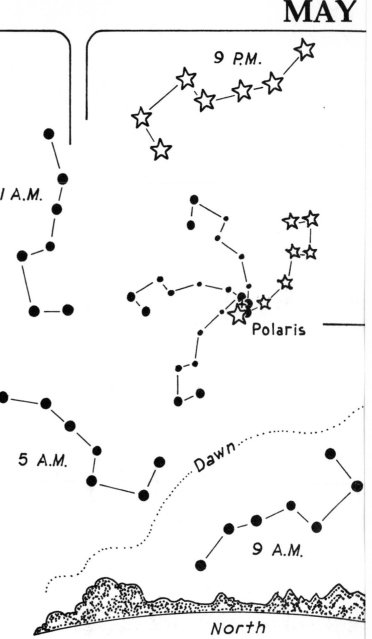

9 P.M.

1 A.M.

5 A.M.

Dawn

9 A.M.

Polaris

North

20th: Polaris takes its name from its place near the **pole.** It is certainly one of the most familiar stars in the sky because of its privileged position. If the axis of the earth were extended it would intersect the celestial sphere very close to Polaris. Like a dot painted on the hub of a wheel, Polaris is the one star that does not seem to move as the earth turns. It is the one star whose place in the sky does not depend upon the time of the day or the day of the year. As such, it has long been of crucial interest to navigators. It would take a book to relate the lore, legends, and symbolism that are associated with this otherwise unnoteworthy star, which just happened to be "in the right place at the right time." In brightness Polaris ranks 49th among stars. It is hundreds of light years away, and must be very luminous to appear so bright at such a distance. Polaris is a yellow giant star, perhaps even now in its death throes.

21st: Watch the Big and Little Dippers throughout the night and you will see them rotate about the **pole.** Of course, it is actually we who are turning, not the stars. But the impression that we are at rest is a powerful one, and it is easy to understand why our knowledge of a spinning earth is only 400 years old. The earth at our feet seems huge and immobile. The stars appear as tiny pinpricks of moving light. It took a bold and courageous act of the ima[gina]tion to reverse these imp[res]sions. We know now [that] Polaris, a typical star, h[as] diameter several thou[sand] times greater than our pl[anet.] The earth is but a speck of [dust] whirling in the cathedral o[f] stars. To be a citizen of a [dust] fleck requires self-confid[ence] and courage—and the bold[ness] to aspire to become worthy [citi-]zens of the cathedral.

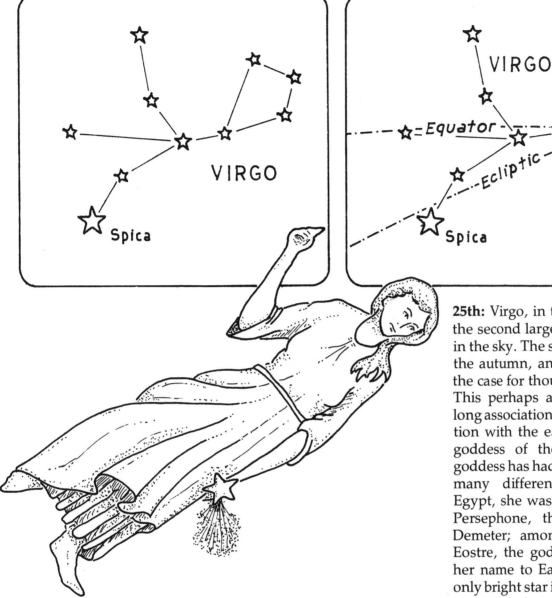

VIRGO

Spica

VIRGO

Equator

Ecliptic

Fall Equinox

Spica

25th: Virgo, in terms of area, is the second largest constellation in the sky. The sun is in Virgo in the autumn, and this has been the case for thousands of years. This perhaps accounts for the long association of the constellation with the earth goddess or goddess of the harvest. The goddess has had many names in many different cultures. In Egypt, she was Isis; in Greece, Persephone, the daughter of Demeter; among the Saxons, Eostre, the goddess who gave her name to Easter. Spica, the only bright star in Virgo, is often represented as a sheaf of wheat which the virgin holds in her left hand. Spica is a giant **main sequence** star, further up the sequence than any other familiarly named star. It is thousands of times more luminous than the sun. Spica lies very near the ecliptic and helps mark that imaginary line in the sky. The planets sometimes wander through this part of the sky, so don't be confused if you find another bright object nearby.

24th: We have been looking toward the north all month. Now, in these last days of May, turn around and face south. The constellation Virgo is halfway up the sky, directly on the celestial equator. There is not much to look for besides Spica (SPY-ka), a dazzling white first-magnitude star. Early in the evening Spica will be almost directly south and halfway between horizon and **zenith.** It has few rivals in this part of the sky. If in doubt, remember how to get there from the handle of the Dipper: "Make an *arc* to *Arc*turus, keep going and you will *spy Spi*ca." It is difficult to associate any of the stars of the constellation with the figure of a virgin. The map above outlines a few of the brightest stars in this inconspicuous constellation. Unless the night is especially bright, it is best to concentrate on Spica alone.

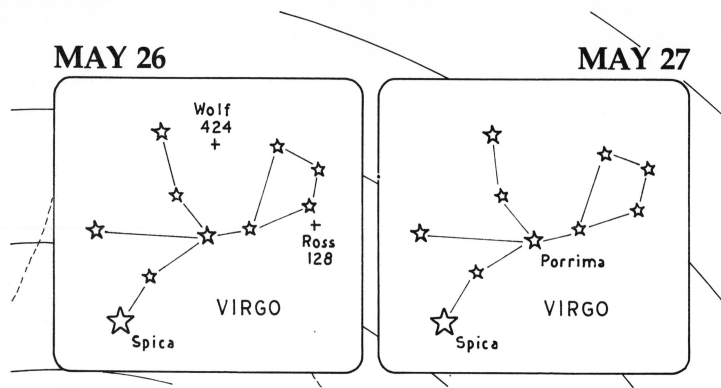

26th: Virgo offers a wonderful opportunity to see deep into the universe, almost to the very edge of space and the beginning of time. We can begin with two of our closest neighbors, Ross 128 and Wolf 424, the 11th and 28th closest star systems to our own. As you might guess, both stars are **red dwarfs** invisible to the naked eye. Ross 128 is the very next star outside of the 10-light-year neighborhood we studied on April 16. It is about 2½ times more distant than our nearest neighbor Alpha Centauri.

27th: Looking deeper into the blackness of the starry night, past ten thousand other stars, we come to Porrima (*PAUR-ih-ma*), named for the Roman goddess of childbirth. Porrima is 35 light years away. A small telescope resolves Porrima into two brilliant white stars of almost equal color and magnitude. It is a binary well worth looking for if you have a telescope, one of the best in the sky. But do it "soon"; the motions of the two stars about each other are bringing them more into line with our line of sight, making it

more difficult to see them as separate stars. Continuing our journey outward, we reach Spica, the 16th brightest star in the sky even at its huge distance of several hundred light years. There are almost certainly at least a hundred thousand stars closer to us than Spica. But we would have to go even deeper into the Milky Way to reach the supergiants Rigel and Betelgeuse. At something near a thousand light years we would reach (in the direction of Virgo) the edge of the galaxy.

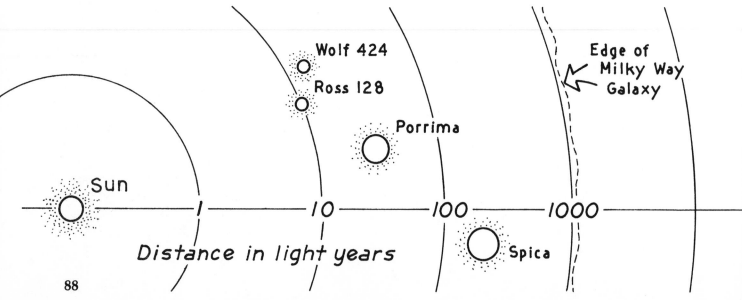

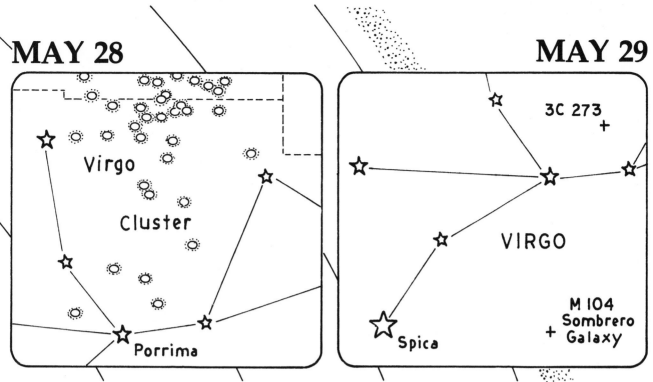

28th: Few parts of the sky are more remarkable than the so-called *field of the nebulae* on the border between the constellations Virgo and Coma Berenices. More than a dozen of the celestial blurs catalogued by Charles Messier lie in this area. We now recognize a giant array of **galaxies** lying 40 or 50 million light years from earth, one of the nearest of many such clusters. It is estimated that there are over 3000 galaxies in this group alone! Each of the galaxies contains tens or hundreds of billions of suns.

29th: If our Milky Way Galaxy were the size of this book, the nearest spiral, the Great Galaxy in Andromeda, would be 17 feet away. The cluster in Virgo would lie 400 feet away. Though it is on the outskirts of the cluster, the Sombrero Galaxy is generally considered part of the group. It is 25 million light years from earth and one of the most marvelous objects to be found in Virgo. The galaxy is splendid on the photographic plate, beautifully symmetric and marked with a broad "sombrero brim" of dark dust lanes. Going

beyond the Virgo cluster, we pass through the realm of the galaxies to the most distant object visible to the owner of a reasonably sized telescope. 3C 273 is the brightest of the **quasars** (3C for *Third Cambridge Catalogue of Radio Sources*). The object lies at the mind-boggling distance of 3 billion light years! We shall discuss the mysterious quasars over the next two starry nights. Continuing our plunge in the direction of Virgo, we would reach, at about 15 billion light years, the edge of the knowable universe.

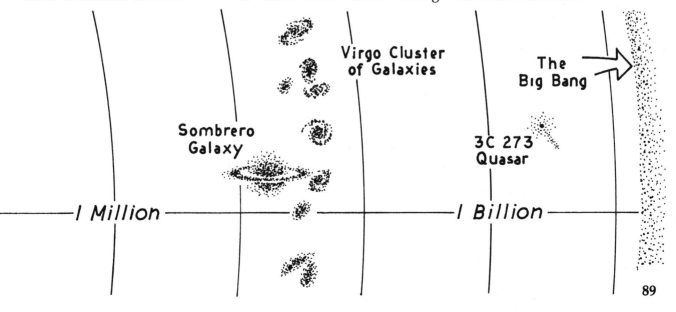

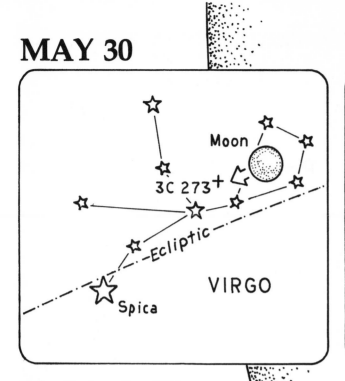

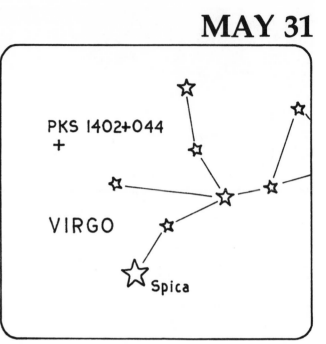

30th: During the 1950s, astronomers discovered sources of radio energy that did not seem to be identified with visible objects. In 1960, a starlike object was discovered near the center of one of these radio sources, 3C 48. The spectrum of the object did not look like that of any known star. It was called a **quasi-stellar radio source,** or **quasar.** Australian radio astronomers were able to pinpoint another of these sources, 3C 273, when the moon, in its journey through Virgo, passed directly over it. Since the moon's position is known with extreme accuracy, it was possible to calculate the location of the moon's edge (limb) at the instant the object was covered and, later, uncovered. The Australians matched the radio source with an inconspicuous 13th-magnitude "star" at the calculated location. In 1962 Maarten Schmidt of the Hale Observatory obtained a spectrum of the "star." The spectrum revealed that the starlike object was billions of light years away. What was it?

31st: Quasars remain a mystery. Over a thousand have been discovered, distributed over the entire sky. For an object to be so bright at such extreme distances, it would have to be more luminous than a hundred galaxies. And yet, quasars appear to be small, probably less than a light year across. That so small an object, more like a star than a galaxy, can emit more energy than a hundred galaxies seems to defy all known laws of physics. Whatever quasars are, they are billions of light years from earth and therefore characteristic of the universe at an earlier stage of its evolution. Also in Virgo is another quasar, PKS 1402+044, at nearly 15 billion light years one of the most distant objects ever observed.

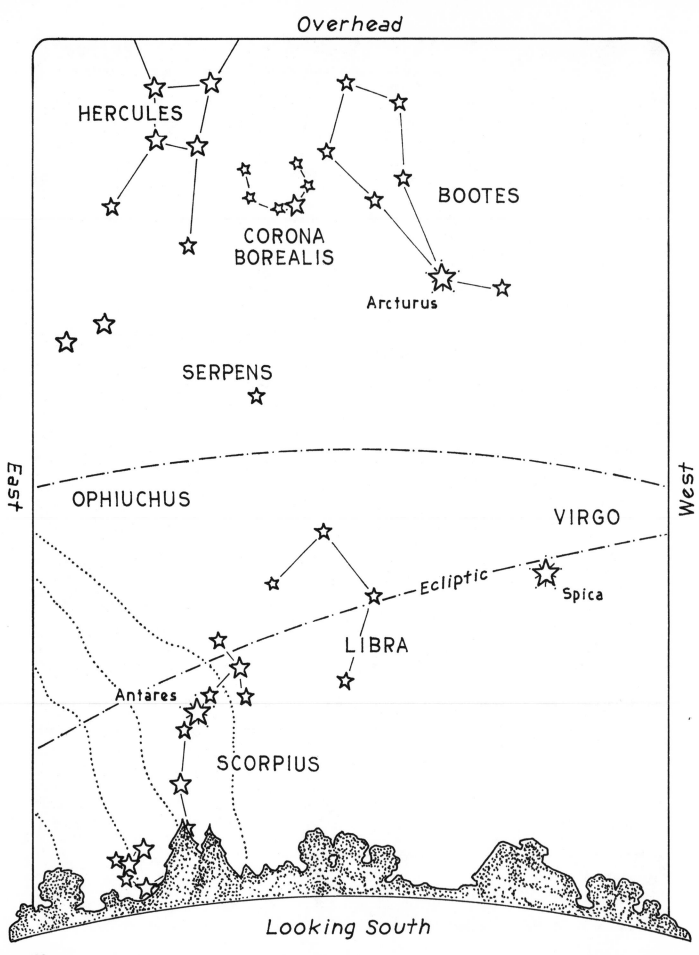

Overhead

HERCULES

CORONA BOREALIS

BOOTES

Arcturus

East

SERPENS

OPHIUCHUS

West

VIRGO

Ecliptic

Spica

LIBRA

Antares

SCORPIUS

Looking South

92

JUNE 1

JUNE 2

Herdsman adapted from Saharan rock painting, 3000 B.C.

CORONA BOREALIS

BOOTES

Arcturus

1st: The north pole of the earth now tilts toward the sun. The sun, in Taurus, nears the top of its northern climb. Today it is 22° above the equator, only 1½° from its northernmost excursion. Stargazers in the southern half of the planet prepare for colder nights. But in the north we feel the warmth of our star's radiant energy falling more directly onto the earth. The daylight hours grow long and we must wait for a later hour for our starry nights. But nights worth waiting for! The brilliant constellations of summer have begun their ascent into our evening skies. And this evening, almost directly over your head, is beautiful Arcturus.

2nd: Five thousand years ago, when the last chill of the Ice Age still lingered in Europe, the Sahara desert in Africa was green and fertile. Late Stone Age herdsmen moved across the plains with newly domesticated cattle, goats, and sheep. And they recorded scenes of the herder's life on the green Sahara in rock paintings of marvelous beauty. Surely, on the spectacular starry nights of those less hurried times, they looked to the heavens and imagined herdsmen among the stars. Bootes *(boo-OH-teez)*, the herdsman, is an ancient constellation. The name is mentioned in Homer's *Odyssey* of the 8th century B.C. In the drawing I have created here, adapted from Saharan rock paintings, I have shown the herdsman of those far-off times driving a cow across the sky. But it is a bear we have before the herdsman in the sky, not a cow. You can find the premier star of Bootes, Arcturus, by following the arc of the bear's tail. The name of the star means "guardian of the bear."

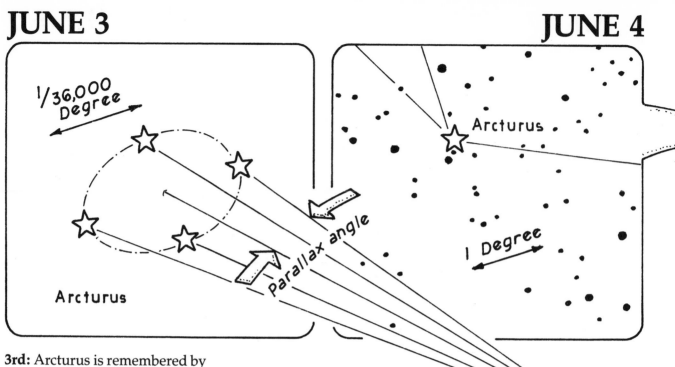

3rd: Arcturus is remembered by many older people as the star that opened the Century of Progress World's Fair in Chicago in 1933. At that time the distance to Arcturus was thought to be 40 light years. There had been a previous world's fair in Chicago in 1893, just 40 years earlier. To establish continuity with that earlier event, light from Arcturus was focused through a telescope onto a photocell which triggered the switches that turned on the lights at the fair grounds. The light that opened the Century of Progress exposition had begun its journey toward earth at the time of the 1893 fair—or so it was thought. Today we know the distance to Arcturus is closer to 36 light years. All of this raises the question of how the distance to the stars is measured. No question can be more basic to our whole understanding of the universe. The method that is used for determining the distance to the nearby stars is outlined on the next few starry nights.

4th: Without knowing the distance to a star, it is impossible to know its true brightness. Without a knowledge of the real brightnesses of stars, the patterns that emerge on an **H-R diagram** would have remained hidden. Those patterns held the key to our understanding of the structure and evolution of stars. The Greeks were the first to measure the distance to the nearest star, the sun. Comparing sun shadows at the ends of a directly measured north-south baseline, they first determined the curvature of the earth and its size. Then, knowing the size of the earth, they carefully observed the moon as it passed through the earth's shadow and worked out the distances to the moon and the sun. The method was refined in later centuries. But it was not until 1833 that the distance to a star other than the sun was measured. The method used by astronomers to determine the distance to nearby stars is the same as that used by surveyors to measure the distance to a landmark on earth. First a baseline is measured. In the case of the stars, it is the largest baseline we can obtain, the 93-million-mile distance from the earth to the sun. As the earth moves around the sun, the position of a nearby star is located against the background of the more distant stars. A nearby star will seem to change position on the celestial sphere when viewed from different places in the earth's orbit. The effect is called **parallax.** From the angular change in a star's position (called the *parallax angle*) and the known length of the baseline, the astronomer can calculate the distance to the star.

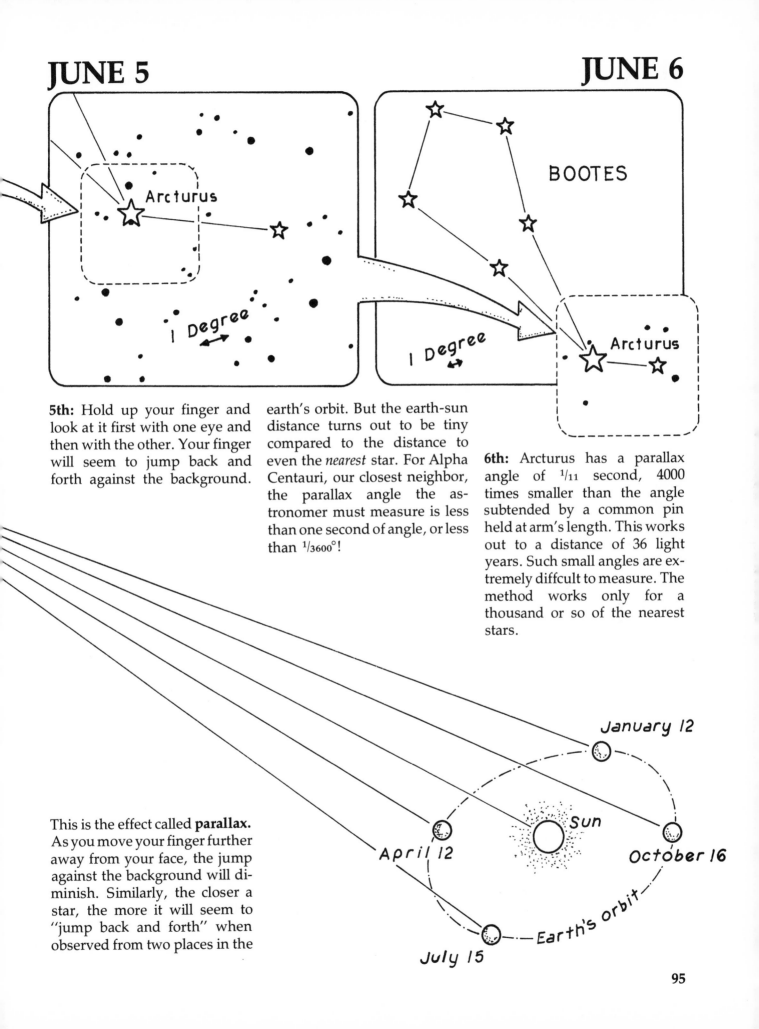

JUNE 5

Arcturus

1 Degree

JUNE 6

BOOTES

Arcturus

1 Degree

5th: Hold up your finger and look at it first with one eye and then with the other. Your finger will seem to jump back and forth against the background.

earth's orbit. But the earth-sun distance turns out to be tiny compared to the distance to even the *nearest* star. For Alpha Centauri, our closest neighbor, the parallax angle the astronomer must measure is less than one second of angle, or less than $1/3600°$!

6th: Arcturus has a parallax angle of $1/11$ second, 4000 times smaller than the angle subtended by a common pin held at arm's length. This works out to a distance of 36 light years. Such small angles are extremely diffcult to measure. The method works only for a thousand or so of the nearest stars.

This is the effect called **parallax.** As you move your finger further away from your face, the jump against the background will diminish. Similarly, the closer a star, the more it will seem to "jump back and forth" when observed from two places in the

January 12

Sun

April 12

October 16

July 15

Earth's orbit

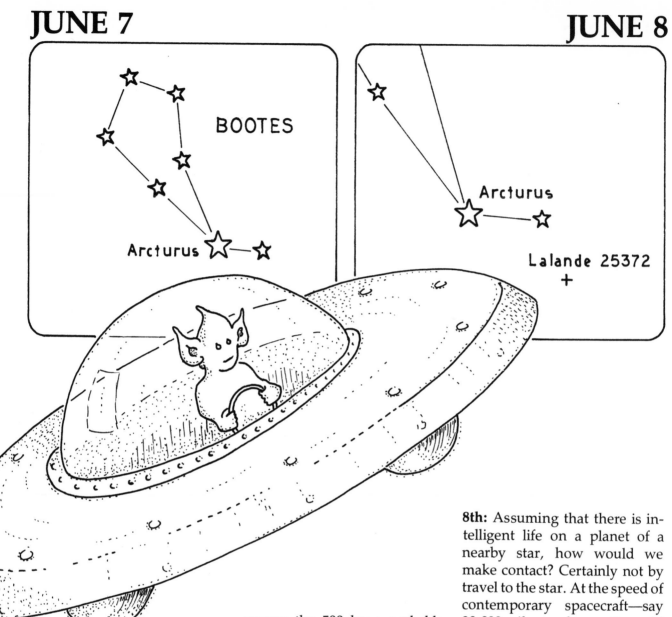

BOOTES

Arcturus

Arcturus

Lalande 25372
+

7th: There are approximately 500 stars closer than Arcturus. What are the chances that there is life on a planet of one of these nearby stars? Probably very slight. Most of the 500 stars are **red dwarfs.** The volume of space around a red dwarf neither too hot nor too cold to support life—*as we know it*—is very small, and therefore unlikely to contain a suitable planet. Hot stars have short lives. The few very hot stars among the 500 have probably not been around long enough for higher life forms to have evolved—assuming billions of years are required. Therefore, our attention is focused on stars in the middle of the **main sequence,** stars like our sun. Of these 500, only 30 or so are stars similar to our sun. Many are binary star systems, complicating the possibility of acceptable planetary orbits. Only a handful of the 500 stars nearer than Arcturus would seem likely homes for fellow citizens of the Milky Way.

8th: Assuming that there is intelligent life on a planet of a nearby star, how would we make contact? Certainly not by travel to the star. At the speed of contemporary spacecraft—say 30,000 miles per hour—it would take almost half a million years to reach Lalande 25372, one of our nearest neighbors. Of course, it is possible that a more advanced civilization has found a way to travel at greater speeds. But a round trip to Arcturus would take 72 years even at the speed of light. It seems unlikely that extraterrestrials would find us interesting enough to have undertaken the thousands of visits described by believers in flying saucers and ancient astronauts.

JUNE 9

JUNE 10

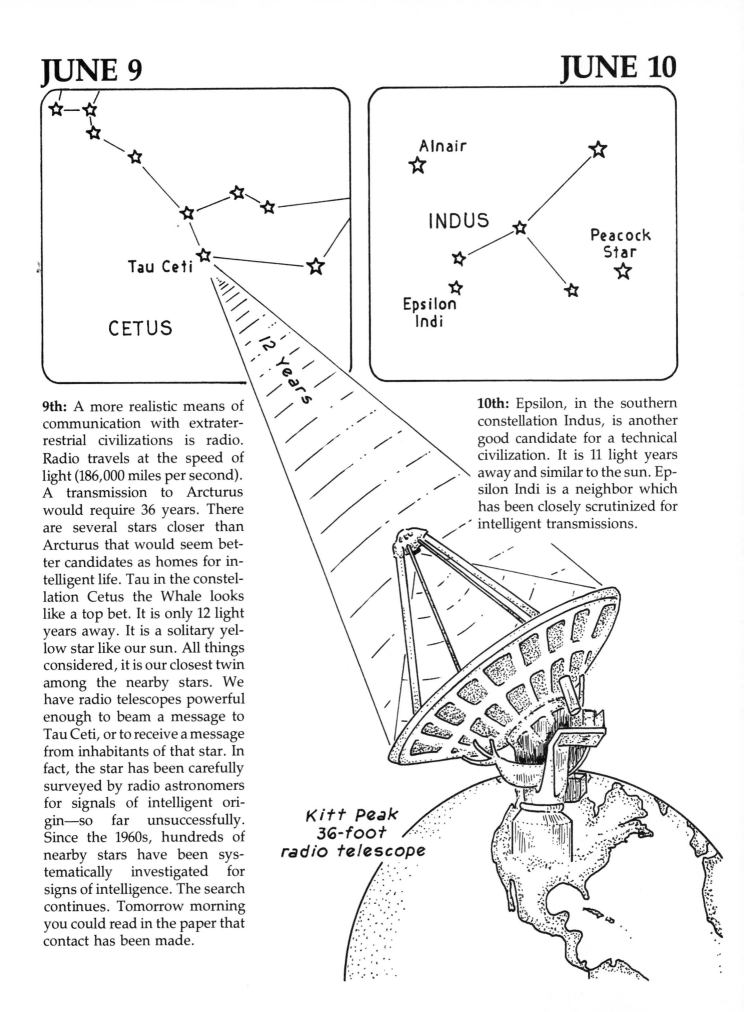

Tau Ceti

CETUS

Alnair

INDUS

Peacock Star

Epsilon Indi

12 Years

Kitt Peak 36-foot radio telescope

9th: A more realistic means of communication with extraterrestrial civilizations is radio. Radio travels at the speed of light (186,000 miles per second). A transmission to Arcturus would require 36 years. There are several stars closer than Arcturus that would seem better candidates as homes for intelligent life. Tau in the constellation Cetus the Whale looks like a top bet. It is only 12 light years away. It is a solitary yellow star like our sun. All things considered, it is our closest twin among the nearby stars. We have radio telescopes powerful enough to beam a message to Tau Ceti, or to receive a message from inhabitants of that star. In fact, the star has been carefully surveyed by radio astronomers for signals of intelligent origin—so far unsuccessfully. Since the 1960s, hundreds of nearby stars have been systematically investigated for signs of intelligence. The search continues. Tomorrow morning you could read in the paper that contact has been made.

10th: Epsilon, in the southern constellation Indus, is another good candidate for a technical civilization. It is 11 light years away and similar to the sun. Epsilon Indi is a neighbor which has been closely scrutinized for intelligent transmissions.

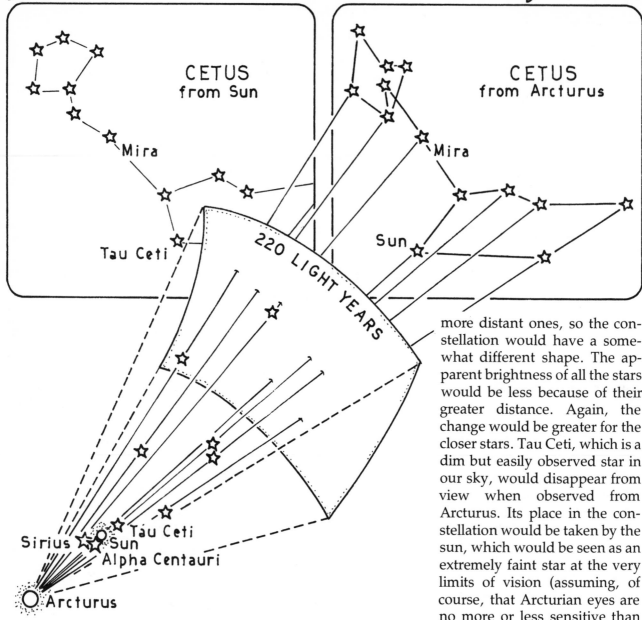

CETUS
from Sun

CETUS
from Arcturus

220 LIGHT YEARS

Mira

Tau Ceti

Sun

Sirius Sun
Tau Ceti
Alpha Centauri

Arcturus

more distant ones, so the constellation would have a somewhat different shape. The apparent brightness of all the stars would be less because of their greater distance. Again, the change would be greater for the closer stars. Tau Ceti, which is a dim but easily observed star in our sky, would disappear from view when observed from Arcturus. Its place in the constellation would be taken by the sun, which would be seen as an extremely faint star at the very limits of vision (assuming, of course, that Arcturian eyes are no more or less sensitive than our own). From Arcturus, Alpha Centauri would be barely visible just to the south of the sun. Our brilliant neighbor Sirius would shrink to a 2nd-magnitude star, but would still be the brightest star in this part of the Arcturian sky, far outshining our sun. Until the time of Copernicus, no one dreamed that the godlike sun was of the same species as the cold points of light in the night sky.

11th: Before we leave Arcturus, let us take a last look from an Arcturian planet back toward the sun. This exercise of the imagination will give us a feel for the way stars are distributed in space. It will also emphasize that the familiar patterns of the constellations depend upon the vantage point from which the stars are viewed.

12th: Tau Ceti in the Whale is one of our nearest neighbors. It is almost exactly opposite Arcturus on the celestial sphere. A citizen of a planet in the Arcturian system, looking our way, would see a very different whale. First, since they are viewed from a greater distance, the stars of the Whale would be a little closer together in the Arcturian sky than in our own. The change would be more apparent for the nearer stars than for the

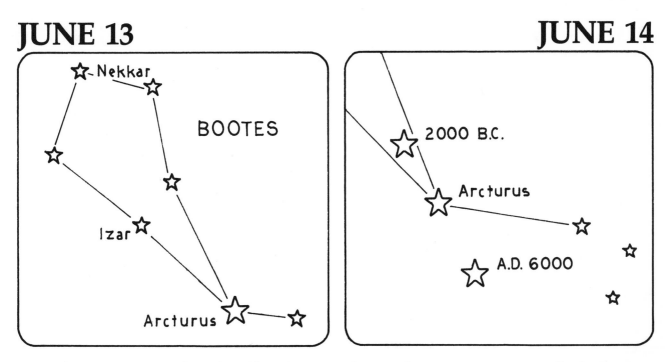

BOOTES

Nekkar

Izar

Arcturus

2000 B.C.

Arcturus

A.D. 6000

13th: Arcturus, the *guardian of the bear,* is the brightest star in the northern hemisphere of the sky, edging out Vega and Capella for that honor. As such, it has been worshipped and feared since earliest times. It may be one of the first stars to have received a name. Arcturus is an orange giant, over twenty times brighter than the sun.

14th: Arcturus has a large **proper motion** and has moved more than a degree since the Greeks began compiling accurate records of the positions of the stars. It was by comparing the modern position of Arcturus with that of the Greek records that the astronomer Edmund Halley first recognized, about 1717, that the so-called fixed stars were not fixed. Arcturus has a motion about the center of our **galaxy** which is steeply inclined to the disk of stars. It is today cutting down through the great spiral disk of the galaxy and will soon pass us. In another half-million years it will have gone its way and faded from sight. Who will then guard the bear?

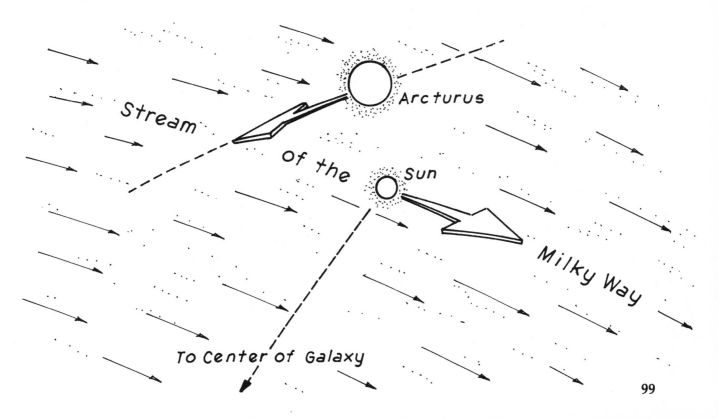

Stream

of the

Arcturus

Sun

Milky Way

To Center of Galaxy

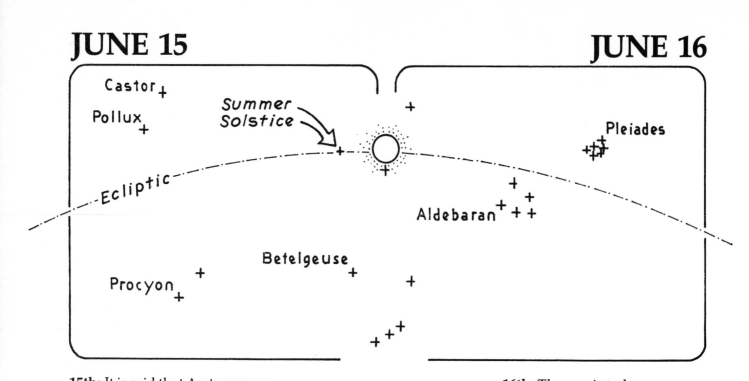

Castor +
Pollux +
Summer Solstice
Ecliptic
Pleiades
Aldebaran + +
Betelgeuse +
Procyon +
+ Rigel

15th: It is said that Arcturus was the first star ever to be observed with a telescope in daylight. This was accomplished in 1635, not long after the invention of the telescope (or its first astronomical use) by Galileo. If you own a telescope, it is fun to go looking for a bright object, say a planet, in daytime. Only an understanding and use of celestial coordinates (see Oct. 9–10) will get you to the sought-for object. I have especially enjoyed seeing the crescent of Venus against the blue daytime sky. We are reminded that the stars are still there during daylight hours. Their faint light is washed out by the overwhelming radiation of our own star, scattered by the earth's atmosphere to fill the sky with pale blue light. The map above shows the sky looking south at noon today. The sun has reached its highest point in its arc from east to west. And there, invisible behind the sun, are our familiar winter stars.

16th: The sun is today very near the Crab Nebula in Taurus (on the map above, its size has been exaggerated). If there was a total eclipse of the sun just at this time, the stars would be briefly visible at noon. The stars we would see would be those we studied on winter starry nights —the stars of Orion, Taurus, the Little Dog, and the Twins— when the earth was halfway around the sun in its orbit through space.

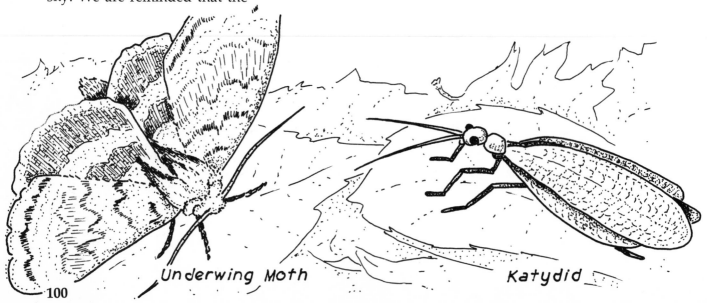

Underwing Moth

Katydid

CORONA BOREALIS

BOOTES

Gemma

Arcturus

Northern Crown

Gemma

18th: The Shawnee story does not end there. The captured sister could not remain forever with mortals, and one night slipped back to her people in the sky. When the sky people saw her husband's grief at his loss, they took pity and brought husband and son to the sky. There the couple resumed their happiness, visiting back and forth between earth and heaven. The brightest star in the circle of dancing sisters is Gemma. Gemma is a second-magnitude star, less bright than nearby Arcturus, but the finest jewel in the crown.

17th: The Greeks called it a *wreath*, the Arabs a *broken dish*. To the Australian aborigines it was the *boomerang*. We know it as Corona Borealis, the Northern Crown, to distinguish it from another tiara of stars in the southern sky. This little circlet of stars, a favorite of star watchers, was long known as the *Crown of Ariadne*. It was Ariadne who saved the Athenian hero Theseus from the minotaur in the maze on Crete. She later married the god Bacchus, who won her love by giving her a magnificent jeweled crown. When she died, after a long and happy life together, Bacchus honored her by putting her

crown among the stars. Another story, of the Shawnee Indians, makes this ring of stars the *Heavenly Sisters* who came down from the sky each night to dance in the fields of earth. There they were spied by a handsome hunter, who fell madly in love with the youngest and prettiest sister. Using his magical powers, the hunter disguised himself as a field mouse and slipped into their circle. When his loved one approached, he returned to his true form and held her tightly as the other sisters scurried back to the sky. When the young girl saw her captor, she immediately loved him. They were married and soon had a son.

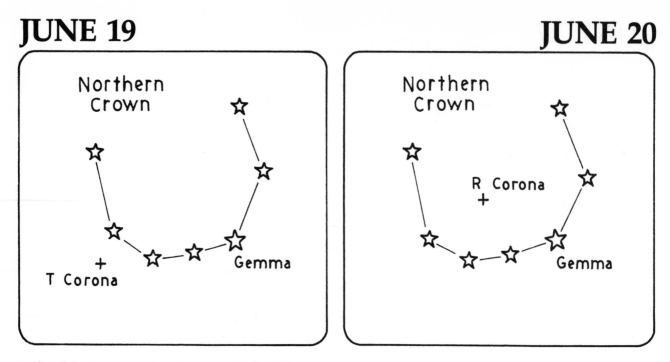

19th: Of the stars in Corona Borealis too faint to be seen with the naked eye, two are of exceptional interest. The first is T Corona, known as the *Blaze Star*. Presently T Corona is an inconspicuous tenth magnitude star, beyond the reach of binoculars. But on at least two occasions in the past, May 12, 1866 and February 9, 1946, the star suddenly flared up to become as bright as Gemma. Such stars are called *recurrent novae*.

20th: The evidence suggests that T Corona is several thousand light years away. To shine as brightly as Gemma at that distance, the star must have been temporarily as luminous as 200,000 suns. On the very night in 1946 when T Corona was ablaze, R Corona was taking a dip. R Corona is an *irregular variable star*. Its usual magnitude is about 6, just below the limit of naked-eye visibility. At irregular intervals the star takes a plunge in brightness. These minima are of irregular and unpredictable duration. Various theories have been put forward to account for the behavior of stars such as these, but none are certain. T Corona and R Corona make us appreciate the steady "life style" of the sun. A blaze or a dip in the sun's radiation could have disastrous consequences for life on planet earth.

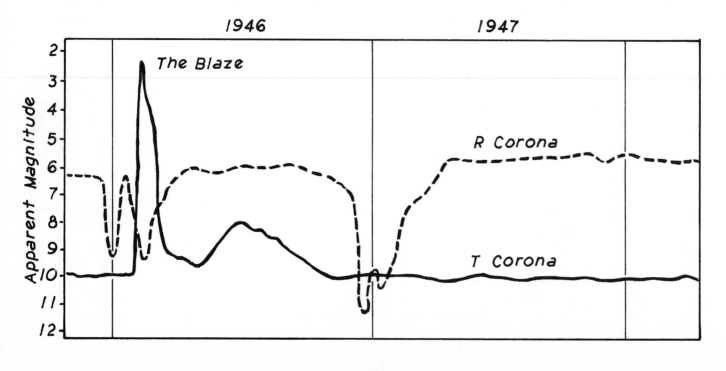

JUNE 21   # JUNE 22

21st: Today is the **summer solstice,** the day on which the northern pole of the earth is tipped the full 23½° toward the sun. As seen from the earth, the sun is as far north in the sky as it ever gets. Now as the earth turns, the sun's journey beneath the horizon is short. Tonight will be the shortest starry night of the year.

22nd: Stargazers above the arctic circle have no starry night at this time of the year. For those far northern observers, the sun never sets. Correspondingly, below the antarctic circle starry nights last 24 hours. It is midsummer in the northern hemisphere; our friends in the south are experiencing winter. We watch the stars in shirtsleeves;

they are wrapped in scarves and sweaters. The word *solstice*, by the way, is derived from the Latin *sol-stitium*, for *sunstanding*. The solstice is the time of the year when the sun stops its northern climb and *stands* briefly before turning back toward the equator.

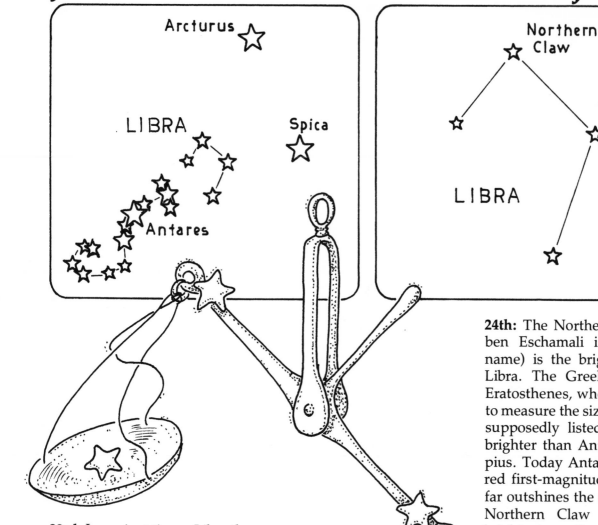

23rd: In ancient times, Libra the Scales were part of the constellation Scorpius. The two brightest stars of Libra are still called *Northern Claw* and *Southern Claw*. It was the Romans who cut the claws off the scorpion and created a new constellation representing the scales of justice. Libra is the only inanimate figure in the zoo of the **zodiac.** The constellation was added to the zodiac at the time Julius Caesar established a new calendar, the Julian calendar. The Julian calendar remained in use until the 16th century. Our present calendar, the Gregorian calendar, was promulgated by Pope Gregory in 1582. Some-times the scales of Libra are associated with nearby Virgo, who plays the role—in this interpretation—of the goddess of justice.

24th: The Northern Claw (Zuben Eschamali is the official name) is the brightest star in Libra. The Greek astronomer Eratosthenes, who was the first to measure the size of the earth, supposedly listed this star as brighter than Antares in Scorpius. Today Antares is a ruby-red first-magnitude star which far outshines the claw. Has the Northern Claw become less bright since Greek times? Or has Antares increased in brightness—perhaps a red giant in the making? No one knows. The Southern Claw, Zuben el Genubi, is only a tiny bit less bright than its companion to the north. No doubt this was the inspiration for the figure of balanced scales. If the balance in the brightness of the stars is as old as the constellation, then perhaps it is Antares that has changed its aspect. Zuben el Genubi, the Southern Claw, lies almost exactly on the **ecliptic,** and is one of those stars that is sometimes **occulted** by the moon. Look for the star halfway between Antares and Spica.

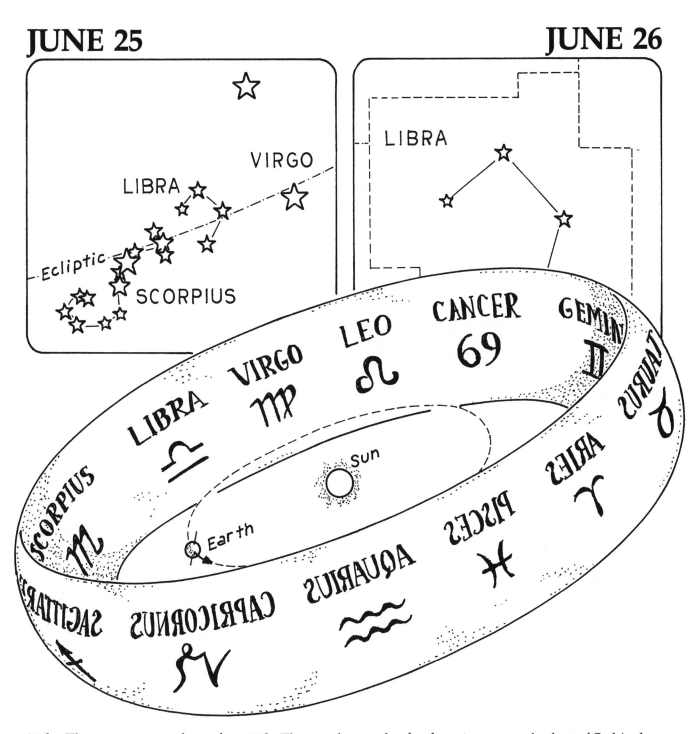

25th: The sun passes through twelve constellations in its apparent journey around the sky, the twelve signs of the zodiac. Libra is the only nonliving figure among them. There are 7½ animals and 4½ humans in the zodiac. The predominance of animals gives us the name *zodiac*—from the same root as *zoo*.

26th: The number twelve for the signs was used as early as 3000 B.C. to correspond to the twelve full months of the year. In fact, the ecliptic passes through thirteen of the official constellations. Ophiuchus, the Serpent Holder, sticks his foot down into that part of the sky between Scorpius and Sagittarius. The sun spends more time journey-ing across the foot of Ophiuchus than it does in Scorpius. The moon will always be found within 5° of the midline of the zodiac (the **ecliptic**). The planets are also confined to the signs, although little Pluto with an orbit highly inclined to the general plane of the solar system can swing 18° away from the ecliptic.

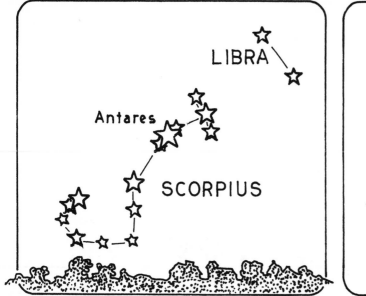

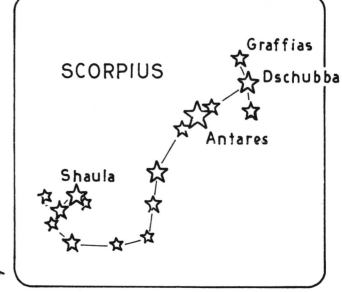

27th: Scorpius the Scorpion is another of those patterns of stars that really suggests the figure it is supposed to represent. In the western tradition, the association of these stars with a scorpion goes back to the very dawn of history. Among the Chinese the long-tailed creature was called a *dragon.* The native cultures of the South Pacific saw a *fishhook* in the dangling curl of stars. Once you have recognized the constellation you will never forget it. You will find Scorpius due south this evening, but very close to the horizon. If there is haze, or the horizon is not open, you will have a hard time seeing it. Look first for the red star Antares, then try to trace out the rest of the constellation. The claws of the scorpion are poised toward Libra. The long tail curls down to the horizon. The whole aspect of the constellation—the blood red color of Antares, the raised sting, the way the creature creeps through the tree-tops—is evil.

28th: Since early times, the Scorpion has represented death, darkness, and evil. Scorpius is the reputed slayer of Orion the Hunter. Orion had boasted of his prowess as a hunter. He would slay, he rashly claimed, all of the animals on earth, without exception. To prevent this calamity, the gods sent Scorpius to sting the giant and put an end to his bloody sport. The resulting poisonous battle caused the death of Orion. The gods put both scorpion and hunter among the stars, but on opposite sides of the sky so they would never fight again. As Scorpius rises in the east, Orion sets in the west. In the drawing, I have shown the scorpion stealing back his claw-stars from the constellation Libra and dashing with his stolen goods back toward the horizon. It is too bad that Scorpius is so low in the sky for northern observers. This is one of the most brilliant star-studded regions of the celestial sphere. The scorpion's tail is hooked deeply into the Milky Way.

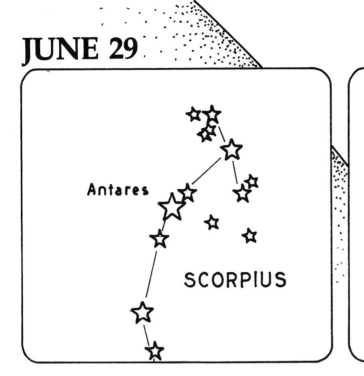

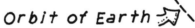

29th: Antares (*an-TARE-eez*) means "rival of Mars" and takes its name from its blood red color. Another name for the star is "heart of the Scorpion." Antares is a supergiant star of monstrous size, altogether suitable for the heart of a monster. Its diameter could be as great as 700 times that of the sun! If the center of Antares was at the sun, its surface would be somewhere near the orbit of Jupiter. Among the familiar stars it is rivaled in size only by huge Betelgeuse. Antares is a bloated star near the end of its life. Because its mass (only 10-15 times that of the sun) is extended over so great a volume, the density of the star is exceedingly small. Like Betelgeuse, Antares is somewhat variable in size and brightness, alternately swelling and shrinking as the delicate balance between gravity and nuclear fusion goes now one way and now another. It is easy to imagine the faintly pulsing red star as the beating heart of the Scorpion.

30th: Shaula (*SHAW-la*), the "sting," makes every list of the 25 brightest stars in the sky, ranking close to Bellatrix in Orion. It is unfamiliar to northern observers because of its position so far south on the celestial sphere. For observers near latitude 40° north, Shaula just skims the treetops. Shaula is the barb on the "fishhook" of the Polynesians. For stargazers in the South Pacific the entire constellation would be far more prominent. At the latitude of New Zealand, Scorpius rides high at the **zenith.**

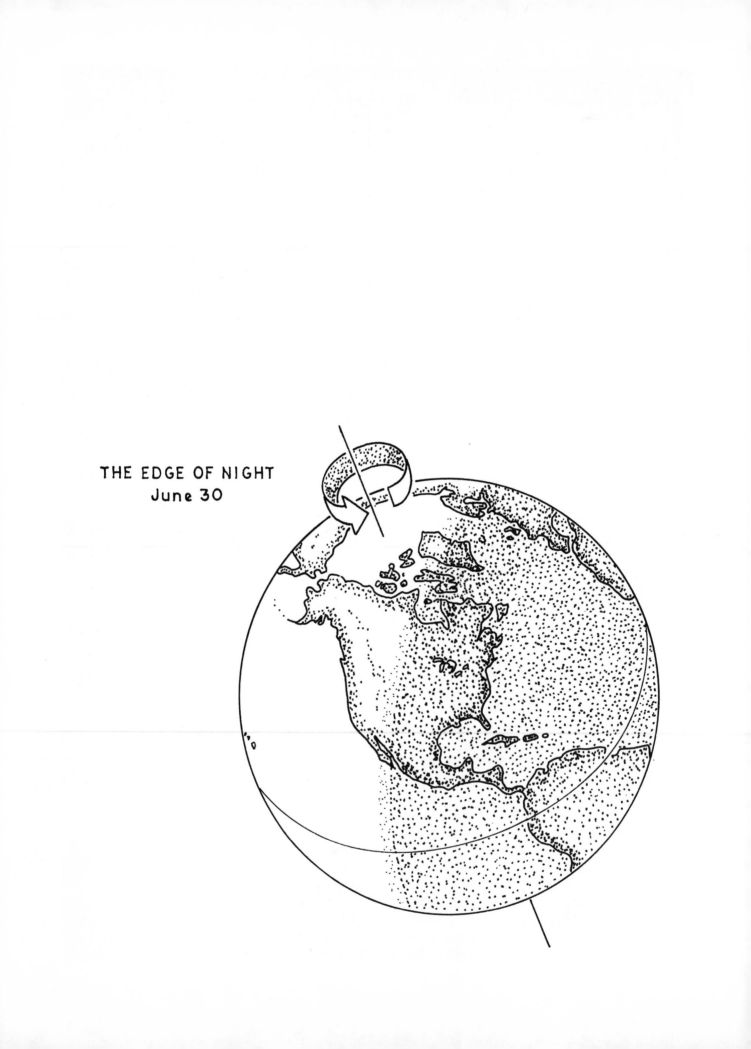

THE EDGE OF NIGHT
June 30

JULY

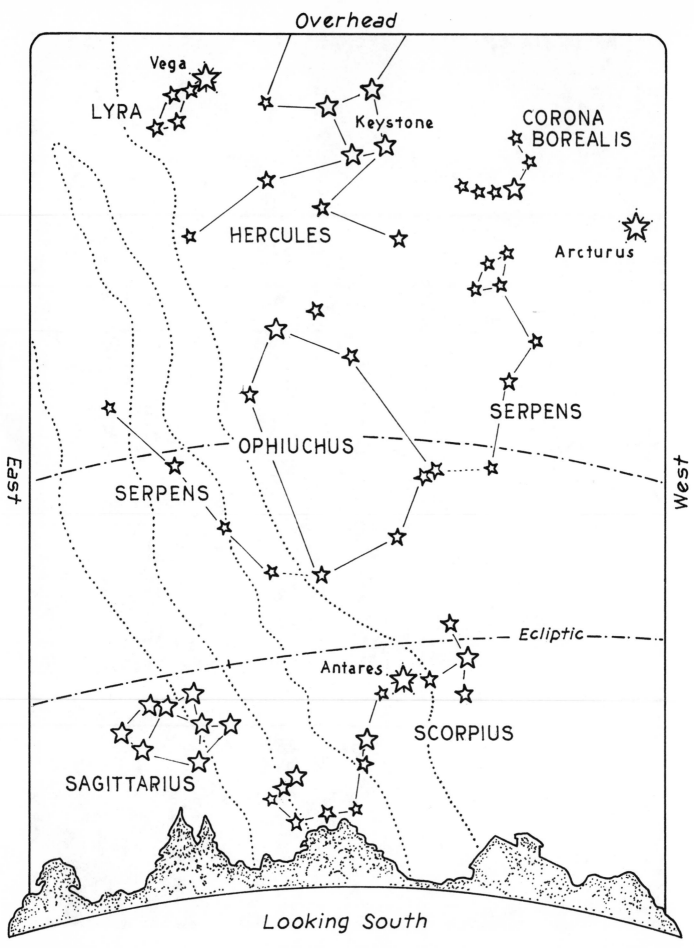

Overhead

Vega

LYRA

Keystone

HERCULES

CORONA
BOREALIS

Arcturus

OPHIUCHUS

SERPENS

SERPENS

East

West

Ecliptic

Antares

SCORPIUS

SAGITTARIUS

Looking South

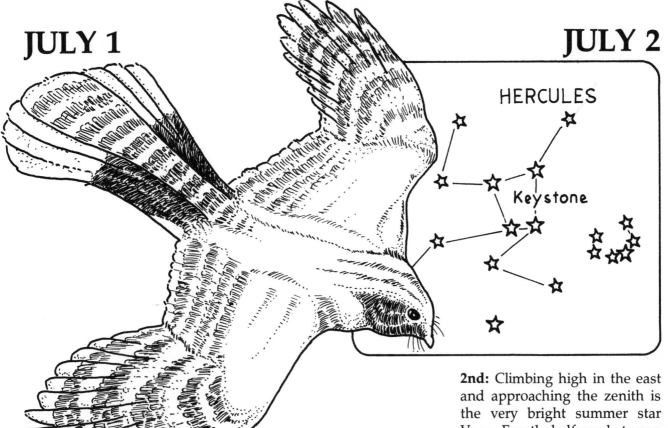

HERCULES

Keystone

1st: No longer the silent nights of winter. Now the starry nights are filled with the sounds of summer. Crickets, cicadas, and night birds add their music to the "music of the spheres." What a beautiful idea was that ancient belief that the stars in their turning make music of a most wonderful consonance. This celestial symphony was, our Greek ancestors believed, too perfect and rarefied for human ears. The idea is at least as old as the Pythagoreans, who believed that all of existence is governed by the laws of musical harmony. The heavens in particular—the sphere of the fixed stars and the individual spheres of the planets, sun, and moon—were as finely tuned as a lyre. This celestial harmony was at the heart of the splendid Greek notion of *cosmos*, a universe in which all parts contrib-

uted their balanced measure to the whole. The Greek concept of the "music of the spheres" had a brief revival at the time of the Renaissance. The great theoretical astronomer Johannes Kepler tried to use the laws of music to explain the planetary motions as described by Copernicus. In this endeavor he did not meet with permanent success, but in the effort he discovered what we now call *Kepler's laws of planetary motion.* If there is music coming from the sky, it remains too fine for human ears. But in summer, stargazing is complemented by the "music of the earth." I have sketched here the whippoor-will, whose song begins at sun-set. Like the stargazer, he becomes active when the sun goes down, hunting (as I have drawn him) for lights in the sky.

2nd: Climbing high in the east and approaching the zenith is the very bright summer star Vega. Exactly halfway between Vega and Arcturus, just to the east of the Northern Crown, lies the constellation Hercules. You can't miss Vega and Arcturus; they are the two brightest stars in the northern hemisphere of the sky. The stars of Hercules, however, are rather incon-spicuous. Look especially for the *Keystone* of four stars in a wedge-shaped quadrilateral. They are in perfect position this evening at the top of the arch of the sky. None of the keystone stars are as bright as Gemma in the Northern Crown, but the overall configuration is easy to find and remember.

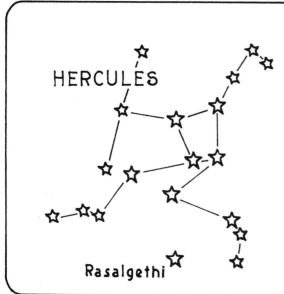

HERCULES

Rasalgethi

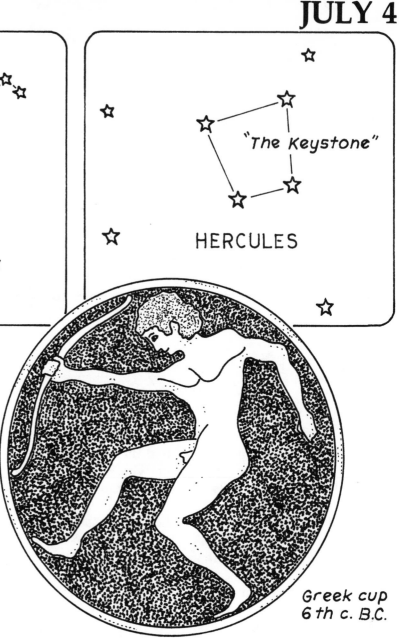

"The Keystone"

HERCULES

Greek cup
6th c. B.C.

3rd: I have shown Hercules here in a rather unconventional posture. The hero is usually depicted the other way round, with his feet to the north and one foot on the head of Draco the Dragon. The name of the star Rasalgethi (*ras-el-GEE-thee*) means "head of the kneeling man." But the *Keystone* reminds me of a torso, so I choose to see the figure with his head to the north, and that is the way I have drawn him. In any case, this is a very ancient constellation, going back, some say, to the Sumerians, and known early in its history as "the kneeling one." We now know the figure as Hercules, the son of Zeus and the mortal woman Alcmene. There are many stories recounting the adventures of Hercules, and many of them have associations with other constellations. Leo the Lion, Hydra the Water Snake, and Cancer the Crab were all creatures defeated by Hercules in the course of completing the famous twelve labors. He is the summer rival of the winter giant Orion.

4th: On the old star maps, Hercules is usually shown with a lion's-skin shield and a club. I have adapted a figure from a Greek cup of the 6th century B.C. and put a bow in his hand to match the lines I have drawn connecting the stars. The stars of the constellation are scattered and faint, and you can make your own attempt to find the figure of a man. I usually concentrate on the *Keystone* and forget the rest. The stars of the Keystone are all of about the third magnitude. If they did not occupy a rather empty region of the celestial sphere, it is unlikely we would give them much attention. There are many binary and variable stars in Hercules, but no star of uniquely compelling interest. But there is one object in the constellation which has attracted the attention of everyone who is interested in the treasures of the sky.

JULY 5

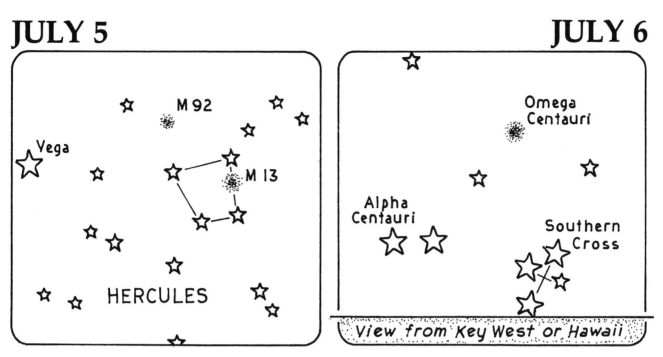

M 92

Vega

M 13

HERCULES

JULY 6

Omega
Centauri

Alpha
Centauri

Southern
Cross

View from Key West or Hawaii

5th: The great jewel of Hercules is the *Great Globular Cluster,* the brightest of these remarkable objects in northern skies. On an exceptionally starry night you might see the cluster as a "fuzzy star" near the limit of vision. Through a fair-sized telescope the "fuzzy star" explodes into a spectacular ball of a thousand stars.

6th: The Great Cluster in Hercules is listed as M 13 in Messier's catalogue of nebulae. A hundred of these ball-shaped clusters have been recognized. **Globular clusters** are distributed about the Milky Way Galaxy like a great spherical halo. A typical cluster is about 150 light years across and contains hundreds of thousands of stars. The brightest globular cluster has a star name, *Omega Centauri* in the southern constellation Centaurus. It is an easy naked-eye object and might be glimpsed from the latitudes of southern Florida or Hawaii, a handspan above the horizon.

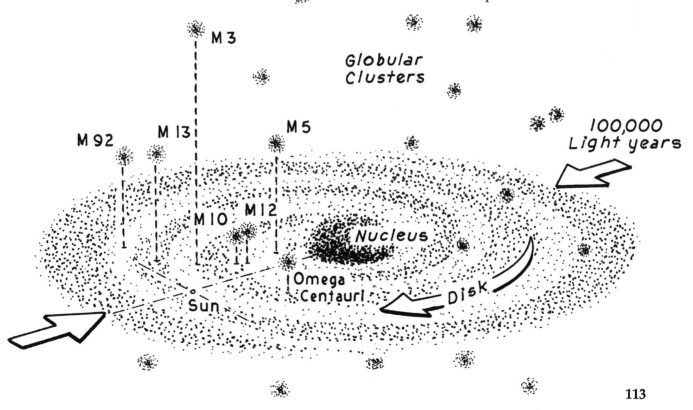

M 3

Globular
Clusters

M 92   M 13   M 5

100,000
Light years

M 10   M 12

Nucleus

Sun

Omega
Centauri   Disk

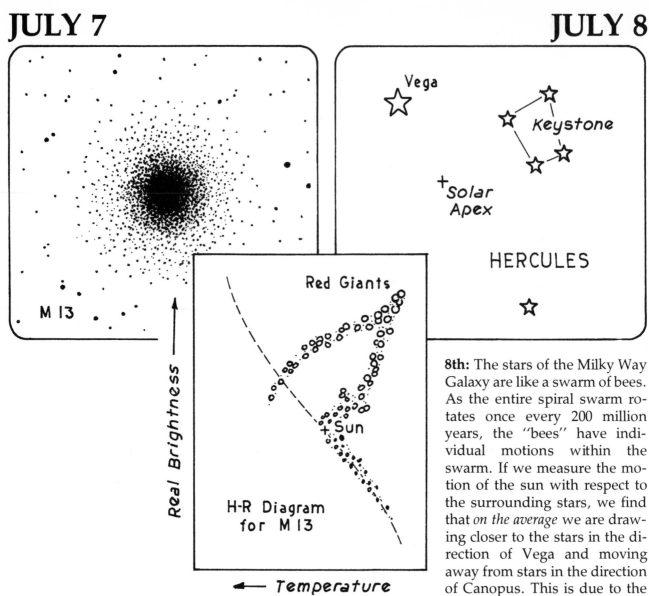

M 13

Red Giants

+ Sun

H-R Diagram
for M 13

Real Brightness ———→

←——— Temperature

Vega

Keystone

+ Solar
Apex

HERCULES

7th: If we construct an **H-R diagram** *(real brightness* vs. *color* or *temperature)* for the stars of the Great Cluster in Hercules, we find that all stars brighter than the sun have left the **main sequence.** Stars with more than one solar mass have become red giants or are on their way to becoming white dwarfs. The theory for energy production in stars suggests that stars such as our sun have sufficient fuel to sustain a normal main sequence existence for 10-15 billion years. It follows that the Great Hercules Cluster must be at least 10 billion years old, or roughly twice as old as the sun. The same observations can be made for all of the other globular clusters. These clusters, then, must be very ancient, probably as old as the **Galaxy** itself. The view from the planet of a star on the outskirts of a globular cluster must be spectacular. One-half the celestial sphere would be ablaze with brilliant red stars. The other half would be lit with the entire spiral sweep of the Milky Way Galaxy. What starry nights!

8th: The stars of the Milky Way Galaxy are like a swarm of bees. As the entire spiral swarm rotates once every 200 million years, the "bees" have individual motions within the swarm. If we measure the motion of the sun with respect to the surrounding stars, we find that *on the average* we are drawing closer to the stars in the direction of Vega and moving away from stars in the direction of Canopus. This is due to the sun's own motion within the "swarm." That point in the sky toward which we are moving is called the *solar apex.* The solar apex is in the constellation Hercules, not far from Vega. Even as we share the almost quarter-billion-year rotation of the galaxy, we move toward Vega at a speed of 12 miles per second. We ride on a merry-go-round (the earth) which rides on a larger merry-go-round (the solar system) which moves on a still greater merry-go-round (the galaxy). Lie on your back under the stars tonight, look out into space, and try to imagine this vertiginous ride!

9th: Ophiuchus *(off-ih-YOU-cus)* is not an easy constellation. There are no first-magnitude stars. The brightest is Rasalhague *(ras-al-HAIG-we)*, which means "head of the serpent holder." The star can be found midway between Arcturus and Altair. Ophiuchus is the man with the snake, and must be considered with the constellation Serpens. I have represented Ophiuchus with a drawing adapted from a Persian bronze figure of the 8th century B.C., showing a god grappling with a monstrous snakelike beast. The struggle of the gods against monsters was an important theme in ancient religions. The struggle was that of light against darkness, of order versus chaos, of good versus evil. It was out of this struggle, according to those ancient faiths, that the ordered universe was created. And it was always possible, if the pressure of the gods relaxed, for the universe to slip back into chaos. It was perhaps of such a faith that Ophiuchus had its origin.

10th: Serpens the Serpent is unique among the constellations in that it is divided into two non-contiguous parts, one on each side of Ophiuchus. The head of the snake is called Serpens Caput, and the tail is Serpens Cauda. The two parts are counted as one constellation. They offer different gifts to the telescopic observer. In Caput is the beautiful globular cluster M 5. Cauda lies near the Milky Way and is rich in star fields, clusters, and nebulae. You will need a vivid imagination to make out the serpent holder *or* the serpent; the array of stars looks to me more like a coffee pot. Ophiuchus is often identified with the god Aesculapius, the first doctor of medicine. The staff of Aesculapius, wound with serpents (the *caduceus*), remains a symbol for medicine.

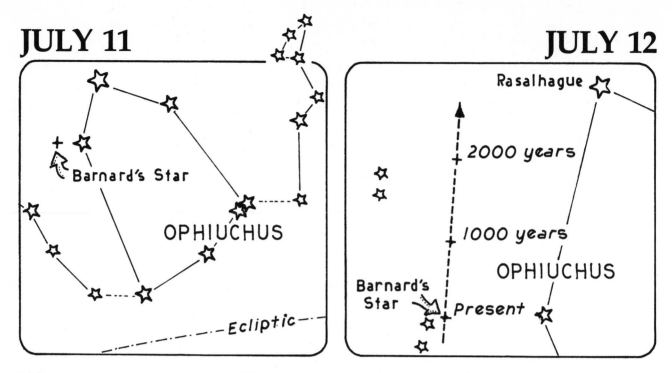

11th: Near the right shoulder of the serpent holder, 6 light years from earth, is our second nearest star neighbor in the universe. Barnard's star is a tiny **red dwarf** discovered by E. Barnard in 1916. It has an apparent magnitude of 9.5, which puts it well below the limit of naked-eye visibility and out of the range of most binoculars. This miniature star is only about 20 times as big as the earth (see Apr. 15–17).

12th: Barnard's star is sometimes called the *Runaway Star.* It has the largest **proper motion** of any known star. It moves a degree across the sky every 350 years, half again as fast as the *Flying Star* in Ursa Major (see May 13). The motion of Barnard's star in three-dimensional space is bringing it ever closer to the sun. Thousands of years from now it will be even closer to us than Alpha Centauri, but still invisible to the eye. Then it

will pass us by and recede into space. From a gentle periodic wobble in the proper motion of Barnard's star it has been deduced that there are possibly two planets, about the sizes of Jupiter and Saturn, orbiting the star. These observations, however, require extremely precise measurements of position over long periods of time. Conclusions deduced from them should be considered tentative.

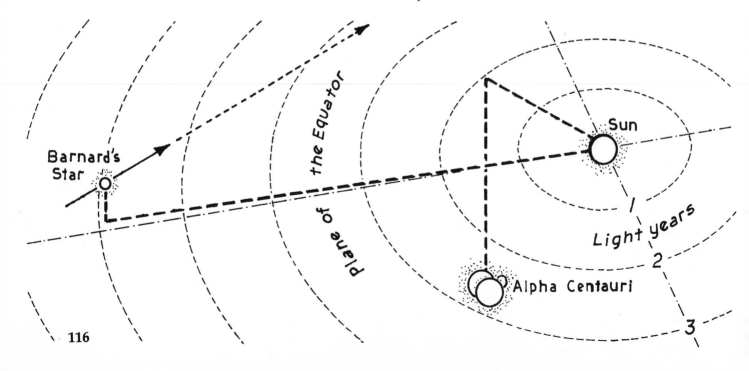

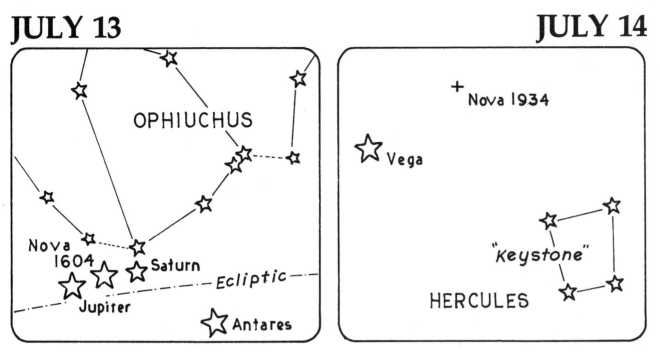

OPHIUCHUS

Nova 1604

Saturn

Ecliptic

Jupiter

Antares

Nova 1934

Vega

"Keystone"

HERCULES

13th: In October 1604, a bright new star flared up in the constellation Ophiuchus. For a time the star was brighter than Jupiter, brighter than any other star in the sky. By wonderful coincidence, the planets Jupiter and Saturn were then very near the place where the nova appeared. The aspect of the three bright bodies must have been spectacular. The **supernova,** as such events are called, remained visible for 18 months before finally fading from sight. The new star was extensively studied by the astronomer Johannes Kepler (see Dec. 27), and has come to be known as *Kepler's Star*. The supernova of 1604 was the last observed supernova in our galaxy. But astronomers have observed these events in other galaxies. As an example, I have illustrated the supernova of 1979 in the beautiful face-on spiral M 100 in Coma Berenices.

14th: Very massive stars, after reaching supergiant status at the end of their lives, can explode in a burst of energy so huge that for some days the star might radiate more light than its entire galaxy. Such an event seems to occur in a galaxy such as the Milky Way once every several hundred years. Kepler's Star was the last in our own galaxy, so we are well overdue for another of these astonishing star deaths. Less massive stars can also flare up at the end of their lives in less catastrophic explosions called **novae,** involving only the outer layers of a star. One of the brightest novae in the 20th century appeared in the constellation Hercules in 1934. A star which had been a 15th-magnitude star suddenly blazed out as a first-magnitude star, and then slowly faded from sight. Novae may sometimes result from the interaction between two stars in a close binary system, and may be recurrent. The theories accounting for these phenomena are still matters for discussion.

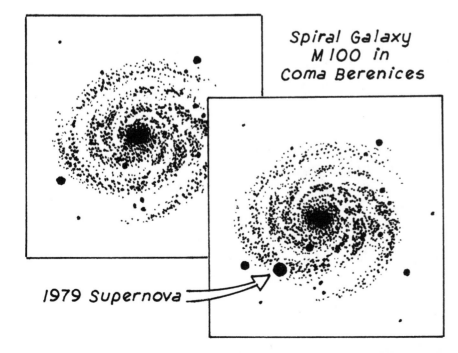

Spiral Galaxy
M 100 in
Coma Berenices

1979 Supernova

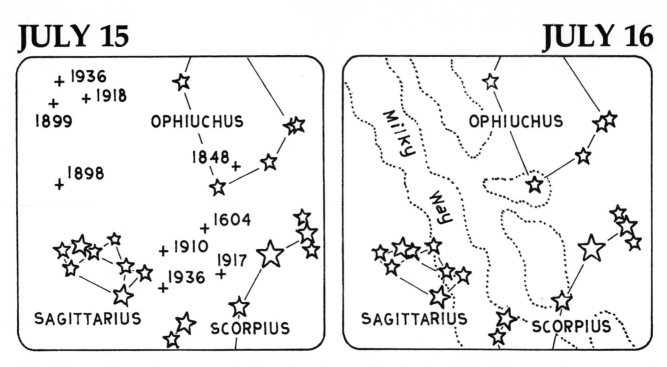

15th: The positions of some recent novae are shown on the star map above. While a **supernova** might occur every few hundred years in a typical galaxy, a **nova** is a much more frequent event. About 200 have been recorded in the Milky Way (only a half-dozen or so supernovae have been observed in historic times). During a flare-up, a nova generates as much energy as the sun does in a million years. The flaring star can brighten by a factor of hundreds of millions of times. The brightest novae in this century appeared in 1901, 1918, 1925, 1934, 1942, and 1975 (see Aug. 24). There is no way to predict when a nova or a supernova will occur. Certainly the next ones we shall see have already occurred, and the great "wavefronts" of intensified light are even now moving toward us.

16th: Novae and **supernovae** usually occur along the band of the Milky Way. The reason is clear. We are able to see these events at great distances, thousands of light years, because of their great brightness. There are far more star candidates for novae or supernovae in the direction of the plane of the galaxy. The sketch below shows the locations in the Galaxy of the three most famous supernovae.

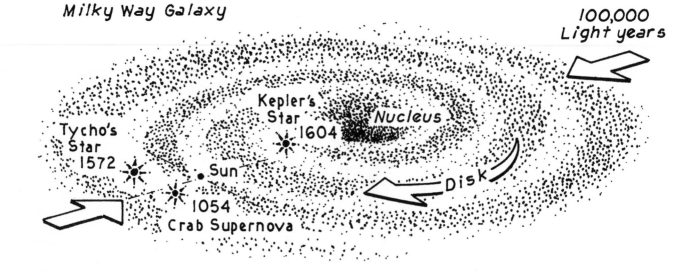

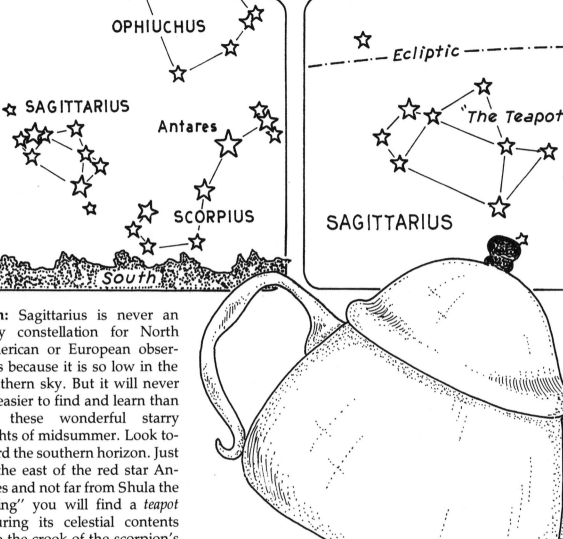

OPHIUCHUS

SAGITTARIUS

Antares

SCORPIUS

South

Ecliptic

"The Teapot"

SAGITTARIUS

17th: Sagittarius is never an easy constellation for North American or European observers because it is so low in the southern sky. But it will never be easier to find and learn than on these wonderful starry nights of midsummer. Look toward the southern horizon. Just to the east of the red star Antares and not far from Shula the "sting" you will find a *teapot* pouring its celestial contents into the crook of the scorpion's tail. There are those who claim to see a bow-bearing centaur among these stars. Sagittarius means "archer" (there is another constellation Sagitta, the "arrow"); you will see him on the next page. But my advice is to stick with the teapot. It may take a while to make the association, but once seen the configuration is easily remembered. Although Sagittarius is an inconspicuous constellation for the naked-eye observer, its hidden delights are many. As we shall see on the next few starry nights, binoculars or telescope reveal an array of wonders.

18th: Sagittarius is one of two centaurs in the sky, a creature half-man, half-horse. He is the only centaur you are likely to see unless you travel south. Sagittarius is supposed to represent the immortal centaur Chiron of Greek mythology. Chiron, we are told, was the mildest and wisest of his rather unruly race. He was much loved by Apollo and Diana (the sun and moon) and became learned in many forms of knowledge from those teachers. In a tragic accident, Chiron was wounded with a poisoned arrow shot from the bow of his own pupil, Hercules. To escape the pain of the wound, the gentle centaur renounced his immortality. At his death he was raised to the sky by Jove. The figure of the centaur, and the myth of his elevation to the sky, are more attractive than the prosaic teapot. But they will not help you find the archer in the sky.

119

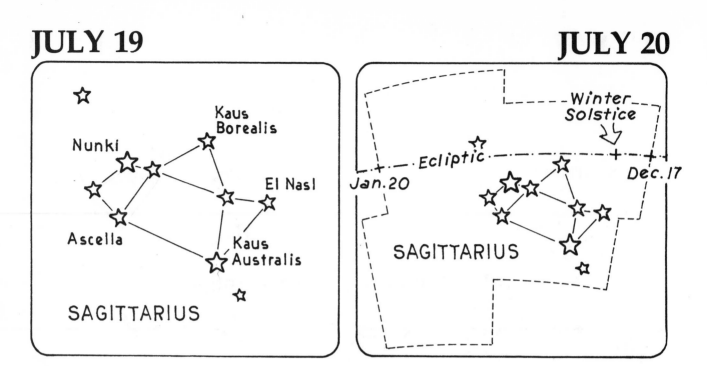

SAGITTARIUS

Kaus Borealis

Nunki

El Nasl

Ascella

Kaus Australis

SAGITTARIUS

Winter Solstice

Ecliptic

Jan. 20

Dec. 17

SAGITTARIUS

19th: Many of the brighter stars of Sagittarius take their names from the figure of the archer. Kaus means "bow," and we have a *northern bow* star and a *southern bow* star. Nasl means "point" and is at the tip of the archer's poised arrow. Ascella refers to the archer's "armpit." The delightful name Nunki has nothing to do with the archer; it refers to the Babylonian god of waters.

20th: Sagittarius is a **zodiac** sign, one of twelve constellations the sun passes through in its circumnavigation of the celestial sphere. The sun is in Sagittarius from December 17 to January 20. It is there that the sun reaches its southernmost excursion. That place among the stars of the archer, and the moment when the sun is there, are called the **winter solstice.** Tonight the sun is nearly half-way around the celestial sphere from Sagittarius, near the border of Gemini and Cancer.

JULY 21

JULY 22

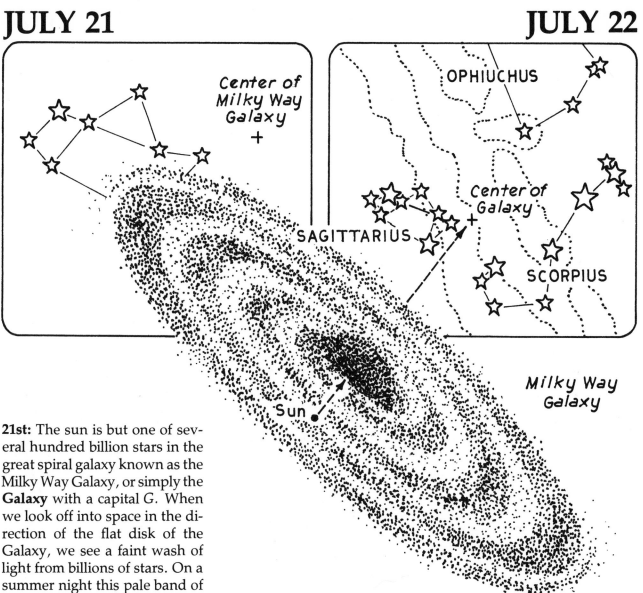

Center of Milky Way Galaxy +

OPHIUCHUS

Center of Galaxy +

SAGITTARIUS

SCORPIUS

Sun

Milky Way Galaxy

21st: The sun is but one of several hundred billion stars in the great spiral galaxy known as the Milky Way Galaxy, or simply the **Galaxy** with a capital *G*. When we look off into space in the direction of the flat disk of the Galaxy, we see a faint wash of light from billions of stars. On a summer night this pale band of diffuse light arches overhead from the northern to southern horizons. It was called **Milky Way** by the ancients, who had no way of knowing its true nature. (Our word *galaxy* comes from the Greek *gala*, which means "milk.") In other myths of antiquity it was known as a celestial river, a bridge linking heaven and earth, or a path to the stars. It wasn't until Galileo turned his telescope to the sky in the winter of 1609–1610 that this flowing stream of milk was resolved into stars, more stars than the eye can count or the imagination conceive.

22nd: The sun's place in the **Galaxy** is about two-thirds of the way out from the center of the spiral. The Galaxy has a diameter of 100,000 light years. We reside 30,000 light years from the nucleus. As we look toward the bright star clouds near the spout of the teapot, we are looking directly toward the nucleus of the Galaxy. The Great Sagittarius Star Cloud, a dense and beautiful part of the Milky Way just north of the spout, is our brief tantalizing hint of the star-filled hub of the spiral. Nowhere else in the sky is the prospect more glorious. But the actual center of the Galaxy is best observed with radio telescopes rather than optical telescopes. So dense are the clouds of dust and gas in the plane of the Galaxy that our view of the central regions is almost totally obscured. Radio telescopes have revealed a powerful source of radio energy at the very heart of the Galaxy. The source appears to be very small, smaller even than our solar system. Some astronomers suspect the existence of a massive **black hole.**

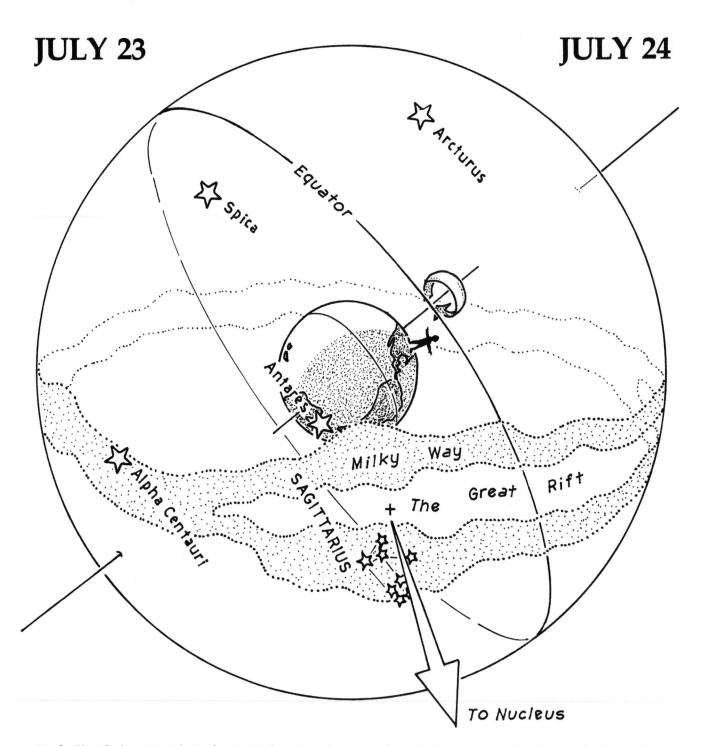

23rd: The **Galaxy** is rich in dust and gas that absorbs the light of distant stars. We see this absorption as a dark gulf dividing that part of the Milky Way which lies toward the center of the Galaxy. This region of absorption is known as the *Great Rift.* The rift divides the stream of the **Milky Way** down the whole arch of the summer sky.

24th: A photograph of the **Milky Way** in Sagittarius is spectacular. We see clouds of stars as thick as salt poured on black cloth. But the Great Rift obscures our view of the very center of the **Galaxy.** What does the dust hide? Our galaxy has a great central bulge, or nucleus, such as we see in the other spirals. In the nucleus the stars are more closely packed, perhaps only fractions of light years apart. But the greater density of stars isn't the only difference between the nucleus of the galaxy and the spiral arms. The nucleus is an extremely powerful source of radio, infrared, and X-ray radiation. The cause of this prodigious outpouring of energy is still a mystery.

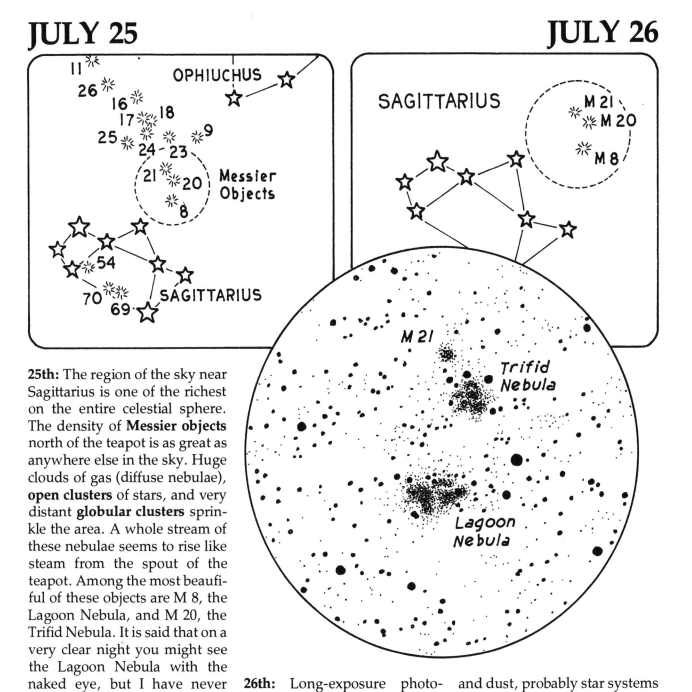

25th: The region of the sky near Sagittarius is one of the richest on the entire celestial sphere. The density of **Messier objects** north of the teapot is as great as anywhere else in the sky. Huge clouds of gas (diffuse nebulae), **open clusters** of stars, and very distant **globular clusters** sprinkle the area. A whole stream of these nebulae seems to rise like steam from the spout of the teapot. Among the most beautiful of these objects are M 8, the Lagoon Nebula, and M 20, the Trifid Nebula. It is said that on a very clear night you might see the Lagoon Nebula with the naked eye, but I have never seen it. Perhaps if you live farther south than I do you will have better luck. It was the invention of photography, and its application to the telescope, that truly revealed these marvelous objects. The telescope expands the area of the human eye, and photography extends the time of perception. Together more of the faint light of the nebulae is registered for our appreciation.

26th: Long-exposure photographs show the *Lagoon Nebula* as an extensive irregular cloud of glowing gas, enveloping a cluster of bright stars. The nebula takes its name from a dusty bay of dark gas that cuts across the cloud. An observatory color photograph of the Lagoon Nebula is, I think, one of the finest artifacts of our civilization. Like the Rosette Nebula in Monoceros (see Feb. 19), the Lagoon has many dark reefs of gas and dust, probably star systems in the process of formation. The nebula lies thousands of light years from earth and is tens of light years in diameter. The *Trifid Nebula* is possibly related to the Lagoon, part of the same corner of the Milky Way. The nebula takes its name from its three-part aspect in a small telescope. A huge O-type star, young and brilliantly hot, is at the center of the Trifid and makes it shine.

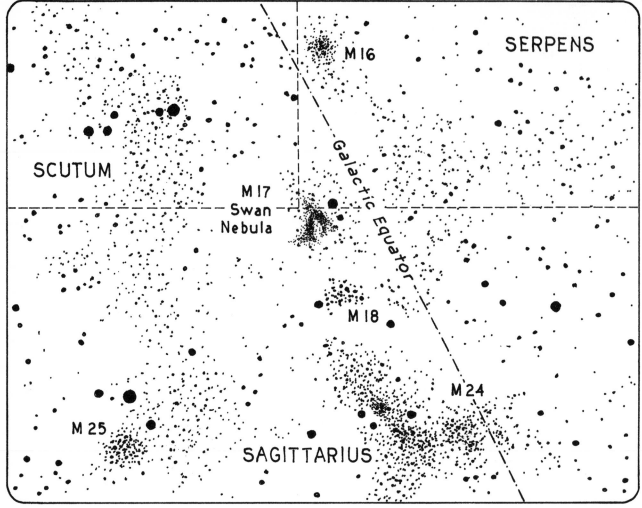

27th: M 17, the *Swan Nebula,* is another of the treasures hidden in the great star clouds of Sagittarius. The blur of light that you might see in a small telescope looks vaguely like a swimming swan. The nebula is also called the Omega or the Horseshoe because of its U-shaped appearance. The Swan is heated and caused to glow by a cluster of stars lying very near the bright core and hidden in the cloud. There is enough matter in this wisp of celestial smoke to form a thousand solar systems such as our own. The distance to the Swan is uncertain, but is certainly thousands of light years.

28th: Charles Messier listed the nebulous objects of the Sagittarius region to prevent them from being mistaken for comets. As celestial phenomena, comets pale to insignificance when compared to the size, complexity, and grandeur of these vast Sagittarian star worlds. Try to imagine thousands of stars, perhaps thousands of "solar systems," tens of thousands of planets, worlds upon worlds upon worlds, all packed into a region only some hundreds of light years in length. This is M 24, the *Small Sagittarius Star Cloud.* The stars are here more closely packed than in the neighborhood of the sun. What glorious starry nights must entertain citizens of worlds in M 24. We can get a hint of the splendor of those nights by scanning the region with binoculars. Wait until a moonless evening when the southern horizon is crystal clear. Look for M 24 about four finger breadths above the knob at the top of the teapot. The entire region will shimmer with the milky light of a million stars. This is the prospect that greeted Galileo when he turned his new telescope to the Milky Way, and realized for the first time the true nature of the stream of light that had (as he said) "vexed philosophers through so many ages."

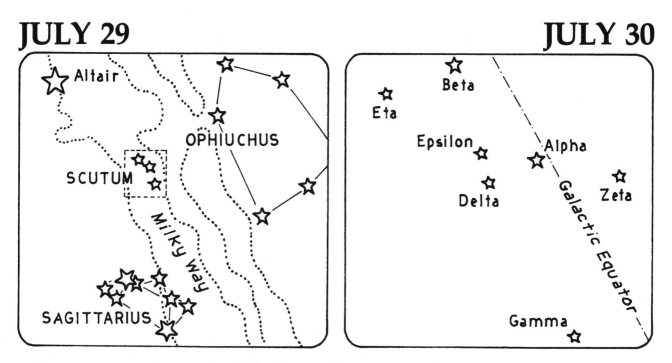

29th: Here is a constellation seldom mentioned in books on the stars. Poor little Scutum, tiny, overlooked, no star brighter than the fourth magnitude. You would need a perfect night to know it was there at all. Tucked between Aquilla and Sagittarius, Scutum is the fifth smallest constellation in the sky. In northern skies, only Sagitta and Equuleus have smaller areas. Scutum *(SKEW-tum)* means *shield*. The constellation's original name was *Sobieski's Shield*, and was invented by the 17th-century astronomer Hevelius in honor of John Sobieski, king of Poland. Sobieski had come to the assistance of Hevelius when the astronomer's observatory in Danzig burned to the ground. It is unfortunate for Sobieski that his name did not survive among the stars. Sextans is another of the constellations named by Hevelius, for a favorite instrument destroyed in that same fire. After centuries of flux, the names of the constellations have been officially fixed since 1930.

30th: Scutum has no named star. The brighter stars are designated by Greek letters, as in other constellations. This is a system introduced by Johan Bayer on the sky maps he created in 1603. The brightest star in a constellation is usually designated *alpha*, the second brightest *beta*, and so on through the alphabet. But the system is far from perfect, and frequently the order of the letters does not remotely correspond to the actual order of brightness. The Big Dipper stars, for example, are lettered from west to east, regardless of brightness. And you would have a hard time finding Alpha and Beta in Sagittarius; there is a whole teapot of stars brighter than these. Nunki, the brightest star in the constellation, is labeled *sigma*!

GREEK ALPHABET

		η eta	π pi
		θ theta	ρ rho
		ι iota	δ sigma
α alpha	κ kappa		τ tau
β beta	λ lambda		υ upsilon
γ gamma	μ mu		ϕ phi
δ delta	ν nu		χ chi
ϵ epsilon	ξ xi		ψ psi
ζ zeta	o omicron		ω omega

JULY 31

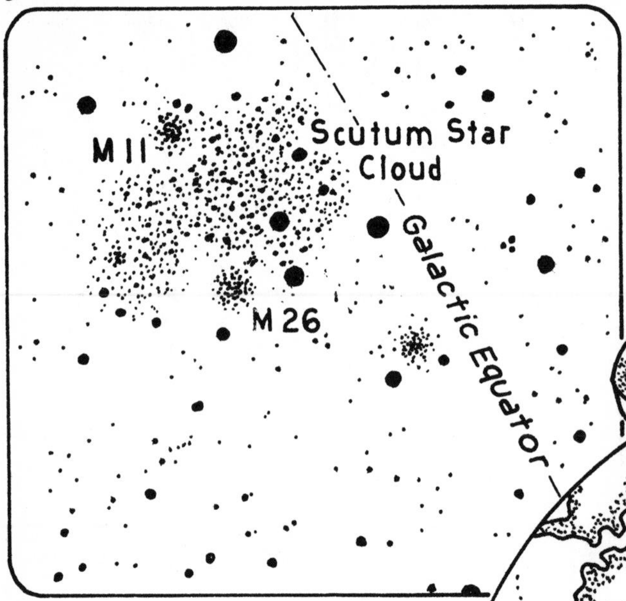

31st: Scutum lies astride the **galactic equator,** the line through the sky that marks the mid-plane of the Milky Way Galaxy. The little constellation is worthy of note if for no other reason than this. The Milky Way is brilliant in this part of the sky, and in few places more splendid than in the Scutum Star Cloud. The Scutum Star Cloud is an incredibly dense mass of stars, bracketed by the rich star clusters M 11 and M 26. The entire area is well worth scanning with binoculars.

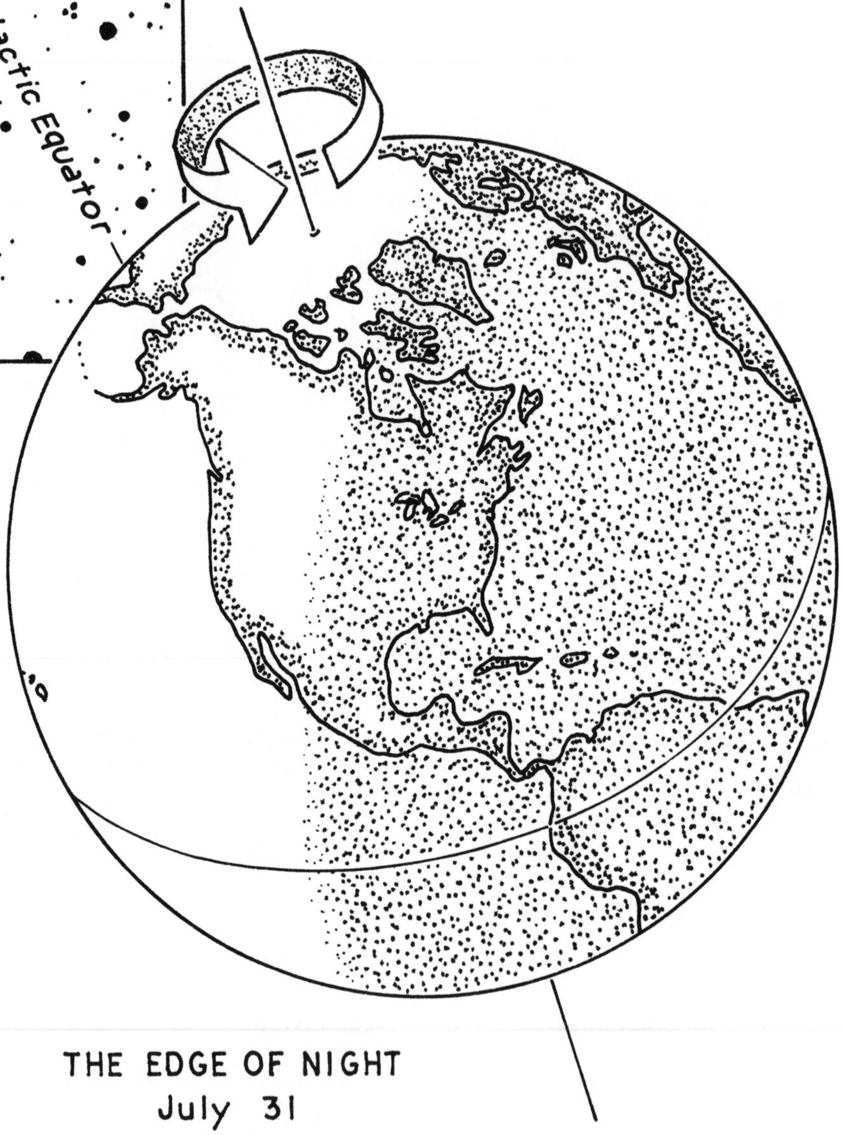

THE EDGE OF NIGHT
July 31

126

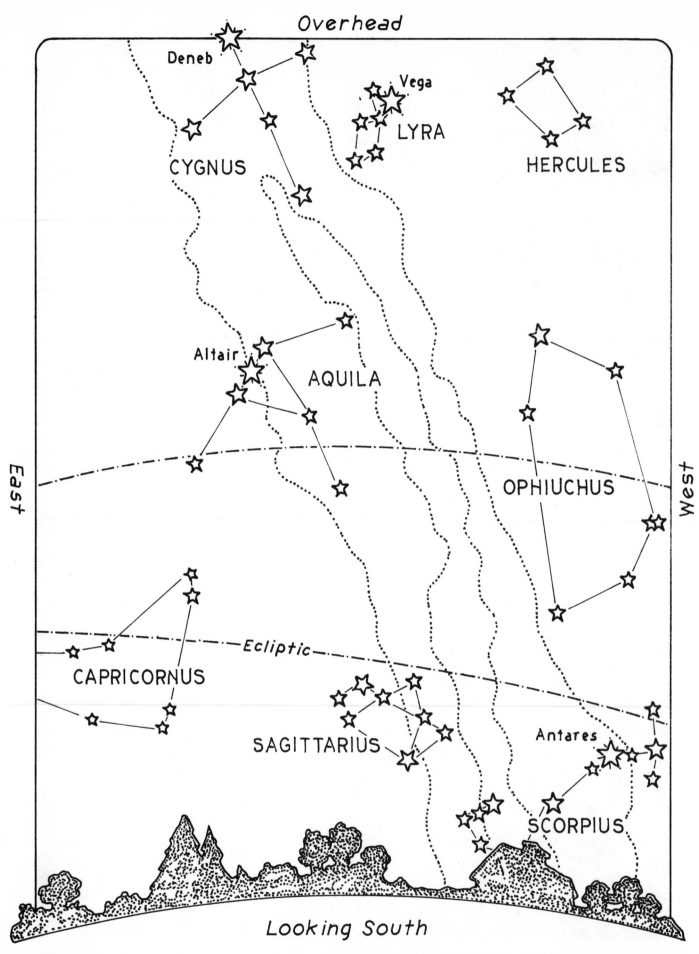

Overhead

Deneb

Vega

LYRA

HERCULES

CYGNUS

Altair

AQUILA

OPHIUCHUS

East

West

Ecliptic

CAPRICORNUS

SAGITTARIUS

Antares

SCORPIUS

Looking South

128

AUGUST 1

1st: The drama of the sky is divided into two acts. Twice each year the Milky Way swings overhead, arching from north to south. With it come the brightest stars and constellations of the year. In winter, Act One brings Orion, Taurus, the Twins, the Big and Little Dogs to center stage. Then there is an intermission as celestial propmen rearrange the set behind the dark curtain of spring skies. And now the second act begins. The brilliant stars of summer—Vega, Deneb, and Altair—make their entrance. If you look overhead this evening, one star almost exactly at the zenith will dominate all others.

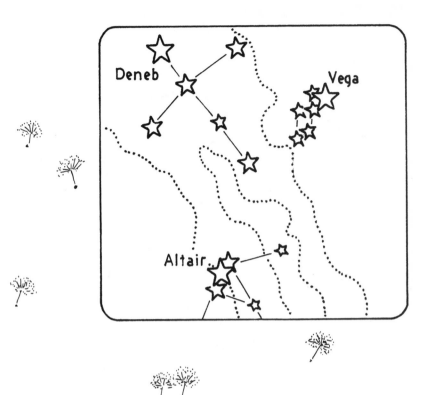

This is Vega, the prima donna of Act Two. Just to the east of Vega, also high overhead, is Deneb in the tail of the swan. Lower in the sky, near the equator is Altair, the third main actor in our drama. Together they form a sky-filling right triangle. This *Summer Triangle* is our best landmark in August skies. If you find it and learn it, you will have no trouble discovering the other things we shall study on these starry nights. August almost certainly provides the best stargazing of the year. The nights are warm, the stars are bright. The most splendid sweep of the Milky Way arches overhead. And the sky is full of shooting stars. No indoor pleasures, no television superspecials, can compare to a night full of shooting stars.

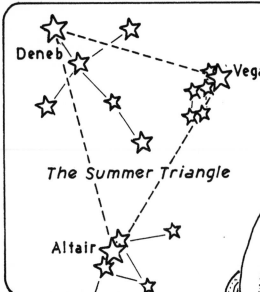

Deneb

Vega

The Summer Triangle

Altair

LYRA

Vega

Figure adapted from Greek cup of about 500 B.C.

2nd: The *Summer Triangle,* your best guide to the sky of late summer and autumn, links three constellations. Like the Big Dipper and the Teapot, it is an unofficial star group, but widely recognized by observers (these unofficial star groups are called *asterisms).* The brightest of the three stars in the Summer Triangle is Vega *(VEE-ga* or *VAY-ga)* in the constellation Lyra *(LYE-ra),* the Lyre. I have illustrated the constellation with a figure adapted from a Greek cup of the 5th or 6th century B.C. The lyre in the hands of the musician is typical of the instruments used at that time. The stars of Lyra do not readily suggest the instrument. Look for a little parallelogram of stars in the direction of Altair. If there were such a thing as the *music of the spheres,* it would likely eminate from the celestial lyre. It is perhaps surprising, given the importance of music in the life of humankind, that we do not find more instruments or musicians among the constellations.

3rd: The lyre was supposedly invented by the god Hermes and given to his half-brother Apollo, who passed it on to his son Orpheus. Orpheus had such a marvelous talent for the magical instrument that, as Shakespeare wrote, "everything that heard him play, even the billows of the sea, hung their heads, and then lay by." It was with his lyre that Orpheus so charmed Pluto and the guardians of the underworld that they allowed his young wife Eurydice, who had died at the bite of a serpent, to return to the world of the living. But Orpheus was warned not to look back to see if Eurydice followed him from Hades until they reached the earth's surface. Unfortunately, his curiosity got the best of him and he lost Eurydice forever. Later, Zeus placed his lyre among the stars—a complement to the wonderful natural music of a summer's night.

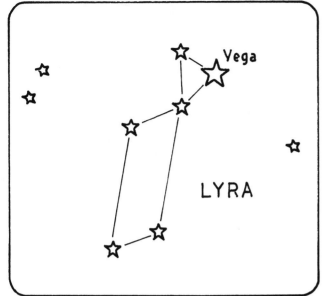

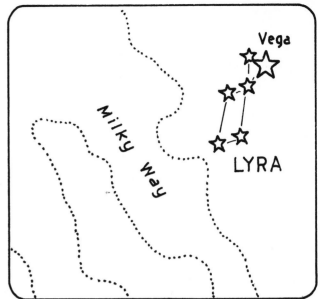

4th: Vega takes its name from the Arabian word for "swooping eagle," a name that goes back to an ancient association of a bird—eagle, vulture, falcon—with this group of stars. Nearby are two other birds, Cygnus the Swan and Aquila the Eagle. The three birds fly the stream of the Milky Way. Vega lies just at the Milky Way's western border.

5th: Vega is a huge blue-white **main sequence** star in the same class as Sirius and Castor. It is about 3 times larger than the sun. Vega is 27 light years away and is the fifth brightest star in the sky, barely edged out of fourth place by Arcturus. Like its winter counterpart Capella, Vega passes almost directly overhead. If you live along the line shown on the map below, Vega will pass exactly through your **zenith** tonight about 3 hours after sunset. At that very moment, Sirius, the brightest star of all, will be more or less under your feet. Down there, hidden by the earth's bulk, the winter stars are rehearsing their next performance.

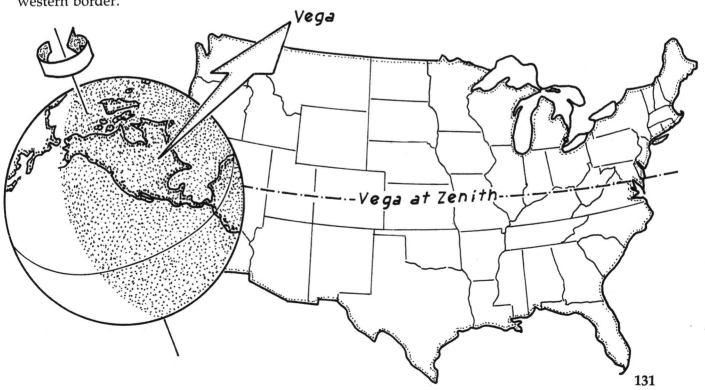

Vega

Epsilon
Vega
Zeta
Delta
Sheliak
Sulaphat

900 LIGHT YEARS

Delta

Sheliak

Sulaphat

Epsilon
Zeta
Vega
Sun

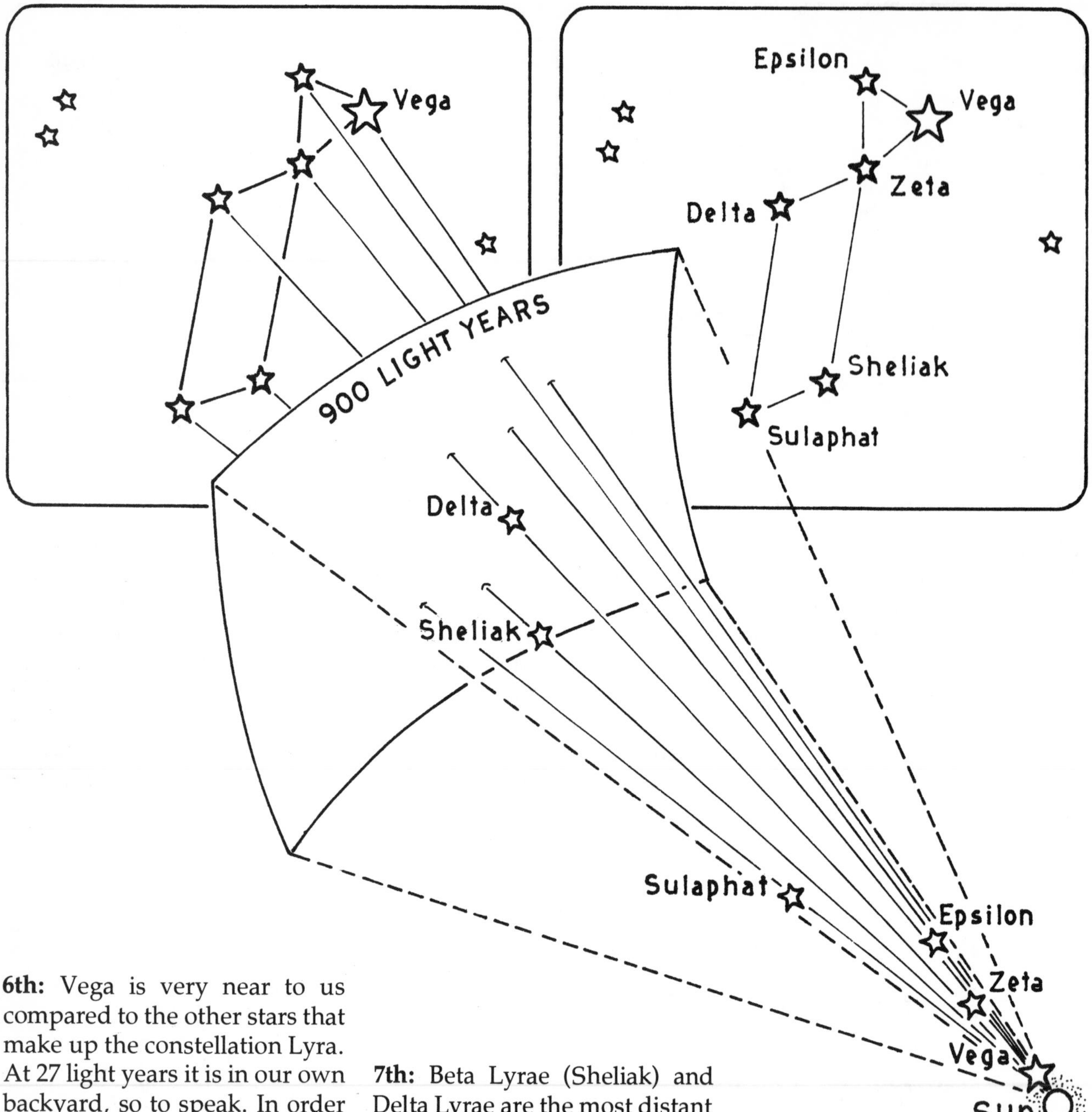

6th: Vega is very near to us compared to the other stars that make up the constellation Lyra. At 27 light years it is in our own backyard, so to speak. In order of distance, a list of the familiar named stars begins *Alpha Centauri, Sirius, Procyon, Altair, Fomalhaut, Vega.* Of course, there are several hundred unnamed and mostly unseen stars closer to us than Vega. Most of these are **red dwarfs.** There are probably many small stars nearer than Vega still waiting to be discovered.

7th: Beta Lyrae (Sheliak) and Delta Lyrae are the most distant stars of the constellation Lyra, lying about 800 light years from earth. The other stars of the familiar six are strung out between here and there. Three of these stars are binary systems (at least) and all of them are more luminous than the sun. The more deeply we peer into space, the more the giant stars tend to dominate as the less bright stars fade from view.

Both Sheliak (*SHELL-yak*) and Sulaphat (*SUE-la-fat*) derive their names from Persian and Arabic words for "tortoise." According to legend, it was from a tortoise shell found on the beach that Hermes fashioned the magical lyre that later came into the hands of Orpheus—and still later into the summer sky.

AUGUST 8

AUGUST 9

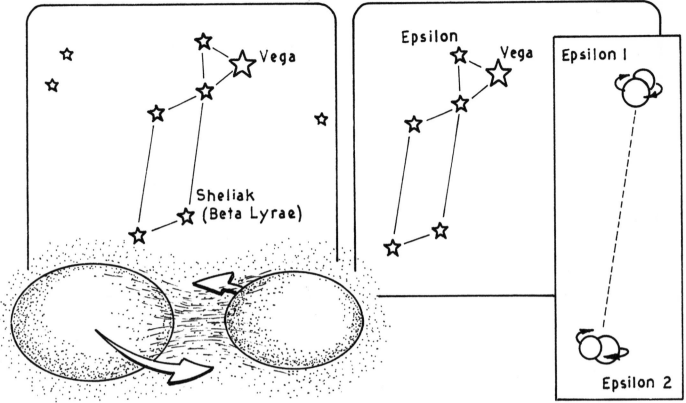

8th: Sheliak (Beta Lyrae) is one of the most puzzling stars in the sky. Its spectra and light variations have been exhaustively analyzed, but are not easily understood. The most widely held interpretation is that Beta Lyrae is a close binary pair, a B-type star and an F-type star. The two stars are so close to each other, presumably, that their shapes are distorted by the rapidity of their rotation and mutual gravitational attraction. As they ro-

tate about each other, each star periodically eclipses its partner, giving rise to the light variations shown below. The stars are perhaps so close that their atmospheres actually intermingle and mass streams back and forth between them. This will result, at some distant time, in the mass of the two stars becoming equal. Beta Lyrae has given its name to the class of close binaries of this type.

9th: Epsilon Lyrae is the famous "double-double" star, a quadruple star system made up of two close binaries. To the eye Epsilon appears as a single star. Binoculars will reveal two stars, a sight well worth looking for. A small telescope will show that each of these components is itself a double, a total of four stars bound together by gravity in an endless dance. All four stars are probably larger than the sun.

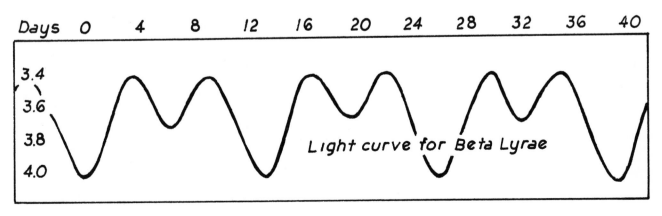

133

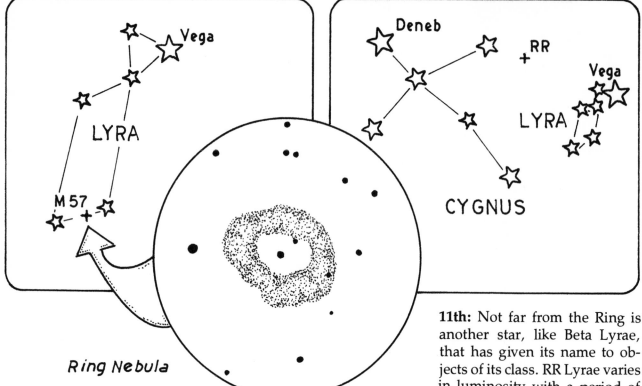

Ring Nebula

10th: M 57, the *Ring Nebula* in Lyra, is a delightful object for the small telescope. It looks like a little cosmic smoke ring adrift in the starry night. The nebula is a doughnut or spherical shell of gas, blown off by a star in its death throes. As recorded on color film this **planetary nebula** (see Feb. 28) is one of the most famous and beautiful objects in the sky. The outer edge of the still-expanding ring glows with red hydrogen radiation. The

body of the ring is yellow. The inner regions are tinted with the green light of ionized oxygen. At the very center is a very hot blue dwarf star, the naked core of the star that blew off its outer layers. The central star is perhaps on its way to becoming a **white dwarf.** Meanwhile, the envelope of glowing gas will continue to expand, adding its star-enriched material to the interstellar medium.

11th: Not far from the Ring is another star, like Beta Lyrae, that has given its name to objects of its class. RR Lyrae varies in luminosity with a period of half a day. Its brightness increases rapidly by about a magnitude, and then slowly drops back to the original level, followed by another flare-up. RR Lyrae stars are thought to be nearing the end of their lives. They have probably shed much of their mass through violent expansions in the **red giant** stage, and are now in a phase of instability on their way to becoming **white dwarfs.** They are actually pulsing in size and energy output.

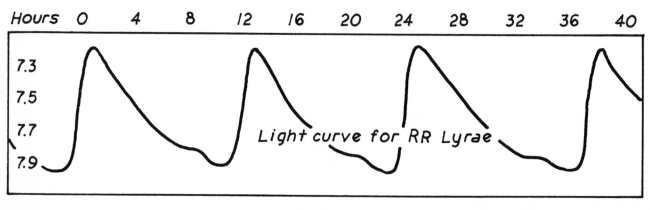

AUGUST 12

AUGUST 13

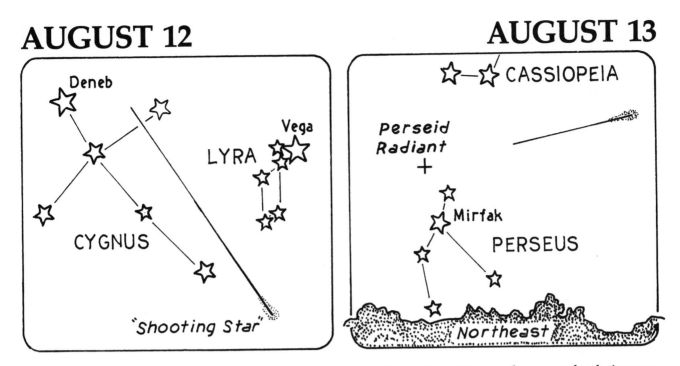

12th: August is the month of "shooting stars." You are more likely to see a **meteor** this starry night than at any other time of the year. Actually, the entire second half of the year is better for meteor watching than the first half. But this is the week that the Perseid shower peaks (see Apr. 11–12), and the "children of Perseus" are the most reliable of all meteor showers.

13th: During August the earth, in its orbit around the sun, passes through the debris of a defunct comet. The material of this comet is now spread out in an elliptical orbit reaching far out into the solar system. As the earth passes through this stream of cometary material, particles enter the earth's atmosphere and are vaporized by friction. We see the heated vapor as a streak of light that looks like a "falling star."

Meteor showers take their name from that part of the sky from which they seem to radiate, in this case the constellation Perseus. Of course, the particles are traveling in parallel orbits around the sun and have no real connection with that constellation. You will be more likely to see a "falling star" in the morning hours, when the earth has turned so that we are facing head-on into the stream.

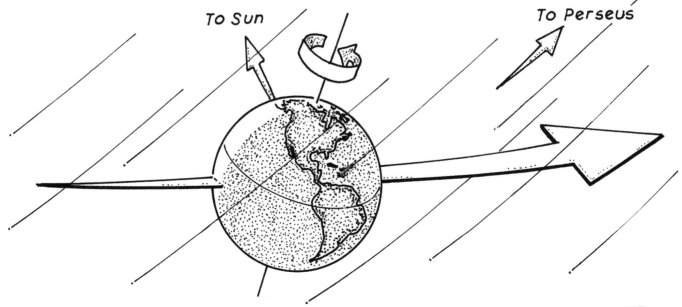

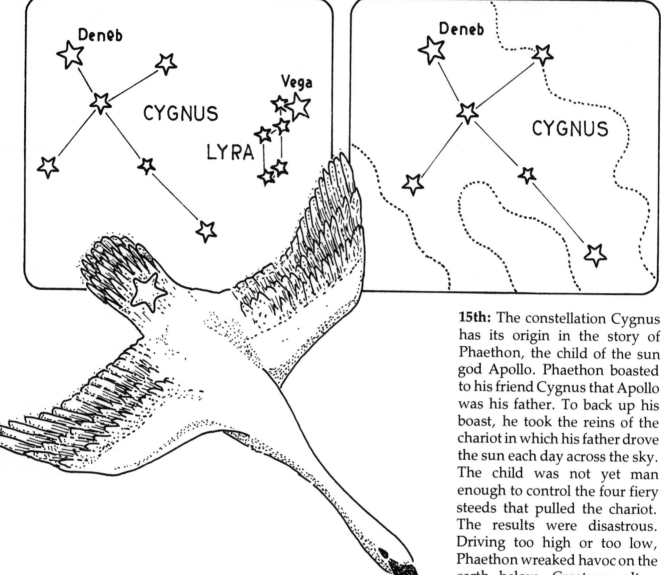

15th: The constellation Cygnus has its origin in the story of Phaethon, the child of the sun god Apollo. Phaethon boasted to his friend Cygnus that Apollo was his father. To back up his boast, he took the reins of the chariot in which his father drove the sun each day across the sky. The child was not yet man enough to control the four fiery steeds that pulled the chariot. The results were disastrous. Driving too high or too low, Phaethon wreaked havoc on the earth below. Creatures alternately froze or fried as Phaethon went careening across the heavens. To stop the chaos caused by the impetuous youth, Zeus struck him dead with a bolt from Olympus. Phaethon's body fell into the river Eridanus. The boy's companion, Cygnus, dove repeatedly into the river in a vain attempt to retrieve the body of his friend. He looked like a swan diving for food. At last, the exhausted Cygnus died of grief. Taking pity, Zeus transformed the boy into a swan and placed him among the stars.

14th: On August nights, high above your head and flying south along the stream of the Milky Way, is Cygnus the Swan. Like other migratory birds, but perhaps a bit too early, the swan wings its way between Vega and Altair toward warmer climes. The night sky is full of birds. Among the 88 official constellations there are nine birds: a swan, an eagle, a crow, a bird of paradise, a dove, a crane, a peacock, the mythical phoenix that rose from the ashes, and a colorful southern toucan. Only the first three members of this celestial aviary are familiar to northern observers; the others fly in southern skies. In addition to the birds, there are several other winged creatures among the stars: Pegasus the flying horse, a fly, and a flying fish. One could perhaps add Sagitta the arrow; after all, it too has feathers and flies.

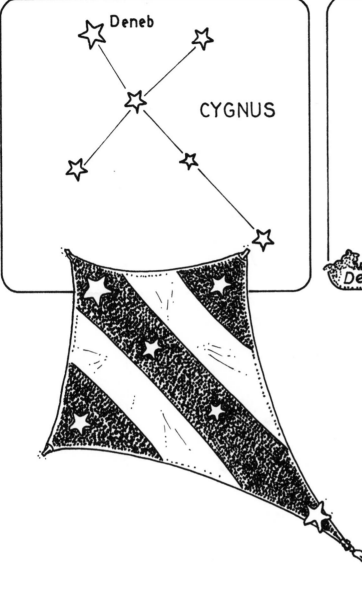

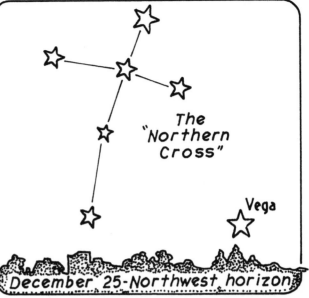

December 25-Northwest horizon

17th: The constellation Cygnus is often called the Northern Cross. There is another cross in southern skies, the constellation officially named "cross" (Crux). The swan is best known as a cross among Christians, and is associated with the cross of Calvary. The symbolism is particularly appropriate around the Christmas season. Then the cross stands upright in the evening on the northwestern horizon. Three factors determine

16th: On a summer's night you can almost hear the beating of the swan's wings as the great bird traverses the **zenith.** Cygnus is another of that handful of constellations that really look like what they are supposed to represent. Once you remember that swans have long necks and short tails, it is easy to find the constellation among the summer stars. The swan's graceful wings are outstretched as it dives between Vega and Altair.

Still, when I am pointing out the constellation, I always suggest that the viewer should look for a kite—an old-fashioned kind of kite, not the new plastic things that come in all shapes. The proportions of a kite are exactly right for Cygnus. However, the name of the brightest star in the constellation fits the swan better than the kite. Deneb (*DEN-ebb*) is the Arabic word for "tail." It is at the top of the kite.

which stars you see in the sky and their orientation to your horizon: your latitude on earth, the time of night, and the day of the year. It takes several seasons of star watching before you get a feeling for the interplay of all three factors. We ride beneath the stars as passengers on a tiny sphere. Some dark night you will suddenly feel, in a way you have never felt before, the roundness of the planet beneath your feet.

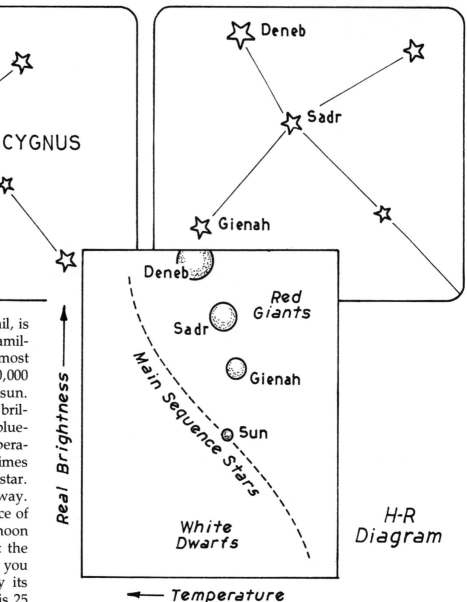

18th: Deneb, the Swan's tail, is unquestioned king of the familiar stars. It is one of the most luminous stars known, 60,000 times brighter than the sun. Only Rigel approaches its brilliance. Deneb is a super blue-white giant, similar in temperature to Sirius but many times larger than that nearby star. Deneb is 1600 light years away. If Deneb was at the distance of Sirius, it would rival the moon in brightness. If it were at the distance of Alpha Centauri you could comfortably read by its light. The mass of Deneb is 25 times that of the sun. When it was still a main sequence star it was at the very top of the sequence. Stars as massive as this remain on the main sequence for only a few million years before exhausting their energy resources and swelling to the status of a **red giant.** Deneb is clearly near the end of its short hot life. The demise of these giant stars is rapid and often catastrophic.

19th: Deneb, Gienah (*gih-EH-nah*), and Sadr (*SAD-er*) are all stars in the final stages of their lives, evolving to or from the **red giant** stage. For stars as massive as Deneb, this takes place quickly, perhaps in a period of tens of thousands of years. Astronomers have studied spectra of Deneb made over the past century in the hope of detecting some change that could be attributed to the evolution of the star. No significant changes have been found. Our ideas about the life cycles of stars are of a theoretical nature (see Mar. 19). The lifetimes of even the most massive stars are long compared to human lives. We have studied the stars in a quantitative way only for a century or two. A century is a snapshot in the life of a star. The sky does, however, offer a family album of snapshots for study. From the album we must reconstruct the family history.

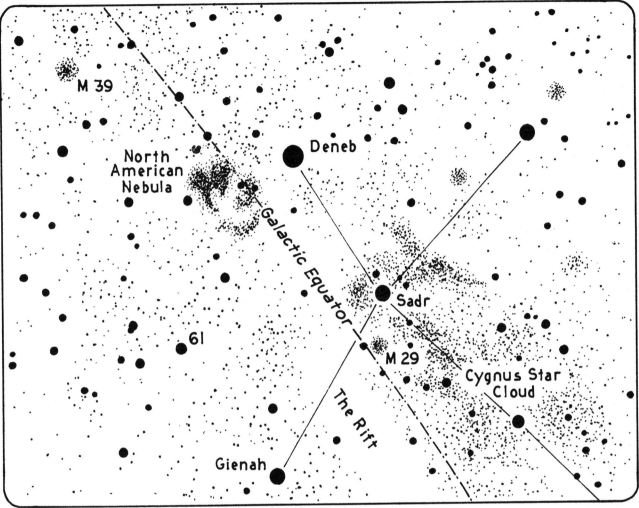

20th: The bright summer Milky Way flows parallel to the body of the swan. The swan's wings arch across the stream of stars. This part of the sky is thick with stars and nebulae. Long-exposure photographs show stars so numerous that their surfaces seem to be almost touching. One would think it impossible for a spaceship to navigate between them. Of course, this is only an illusion; the stars are separated by light years. In this river of stars are many well-defined clusters. Two of these clusters are listed in **Messier's catalogue.** And weaving in and out among the stars are wisps and filaments of glowing gas.

21st: If you have binoculars, scan the Swan. You will be dazzled by the number of tiny stars that feather its body. Scan especially the region near Sadr (Gamma Cygni). Here the stars are clumped in "clouds," and photographs show the clouds overlaid with swirls of gas. The gas is heated to incandescence by the energy of huge yellow Sadr. The star and its attendant nebulae are about 750 light years away. Below Sadr, the *Cygnus Star Cloud* marks one of the spiral arms of our galaxy. The sun is located just at the inner edge of this arm, which is known as the *Cygnus Arm* (see Dec. 11). The stars we see in the Cygnus Star Cloud are thousands of light years distant. Also near Sadr begins the *Great Rift,* a thick band of obscuring dust and gas that divides the Milky Way from here to Scorpius. Only a little imagination is required to see the whole spiraling whirlpool of the Galaxy. Stargazing is, after all, as much imagination as actual perception. This has been so since the days when our ancestors saw swans and lyres in the heavens. It remains so today. You may see a circular blur of light when you look through a telescope at the Ring Nebula. In the mind's eye you will imagine the colossal death of an ancient Deneb.

139

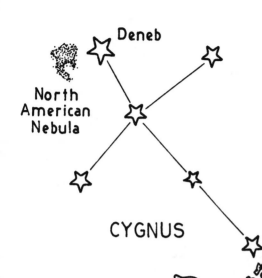

22nd: Just to the east of Deneb lies the beautiful *North American Nebula.* To the naked eye on the clearest night the nebula is no more than an indistinguishable part of that glow of light we call the Milky Way. With binoculars, it becomes a slightly brighter patch of light. On the Palomar color photographic plate the nebula is revealed as an entire continent of stars and interstellar gas waiting to be explored. The nebula was first photographed in 1890, and took its name from its shape. The geographic illusion is created by clouds of glowing gas (an electric pink in the photographs), obscured at the margins by darker masses of opaque matter. Off the east coast, in what should be the Atlantic Ocean, are other clouds and wisps of nebulosity. The largest of these is known as the *Pelican Nebula,* again because of its shape. I like to think of the Pelican as lost Atlantis risen from the sea. As I look at a photograph of the North American Nebula, I am reminded of the explorers who, on horseback and in birchbark canoe, mapped North America. Our own small planet has now been thoroughly explored. We have reached the heart of every dark continent. We now turn our curiosity to new "dark continents" in space and the telescope will be our "birchbark canoe." Perhaps a young reader of this book will become the Lewis or Clark who will map the deepest reaches of the North American Nebula.

23rd: The interstellar nebulae still offer many mysteries to the observational and theoretical astronomer. Although the glowing gas of the North American Nebula appears on a photograph as thick as a summer's cloud in the earth's atmosphere, it is actually exceedingly more rarefied. On the average there may be only a dozen hydrogen atoms per cubic centimeter. On earth, this would be considered a nearly perfect vacuum. By the standard of space between the stars, the gas is thick. It has been estimated that in the huge volume of the nebula there is enough hydrogen to make a hundred suns. For a long time it was thought that the material of the nebula was excited to luminosity by the radiation of brilliant Deneb. That star and the nebula are apparently at about the same distance from earth. But it is now thought unlikely that a star so far removed laterally could have this effect. The source of the nebula's light is probably embedded in the "continent" itself.

140

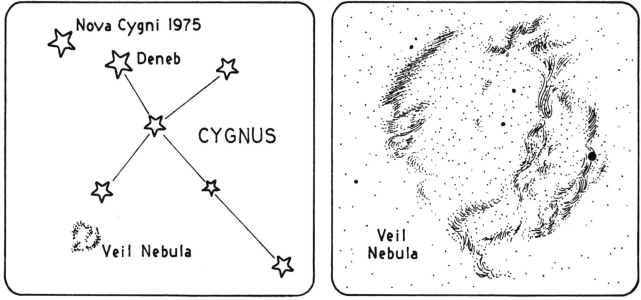

24th: One weekend at the end of August in 1975, stargazers were treated to a rare and beautiful sight. I heard about it when I had a telephone call from a friend, who had just heard a bulletin on the radio. A new star had suddenly and unexpectedly appeared in the constellation Cygnus, just to the northeast of Deneb and not far from the North American Nebula. The star became briefly almost as bright as Deneb, and dramatically changed the appearance of the summer swan. The brightness then began to quickly fade, and by the following weekend the new star, or **nova,** was invisible to the naked eye. The nova was not, of course, a "new star," except for earth observers. The old star that had suddenly flared up in this spectacular way was thousands of light years distant, all the way on the other side of the Cygnus Arm of our Galaxy (see Dec. 11).

25th: Cygnus was the scene of an even more spectacular **supernova** explosion 30,000 years ago. What our prehistoric ancestors made of the dazzling new star, we do not know—but that the event occurred seems certain. Today we see a huge loop of glowing gas still rushing outward from the colossal blast. The delicate filigree of star debris is tinted red, white, and blue in the Palomar photographs. The delicacy of the bunting gives the nebula its name—the *Veil*. The expanding shockwave sweeps up dust and gas as it goes, and has partly "cleaned out" a bubble of space. We can see more faint background stars through the bubble than in the dustier regions outside. The loop is now 70 light years in diameter. There must be stars inside the bubble in addition to the one that exploded. It is interesting to guess at the effects on a planet of one of these stars as the shock wave went by. Has our own planet been the recipient of such a blow at some time in the past? If so, how was life affected?

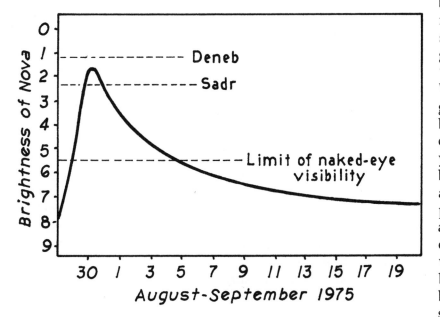

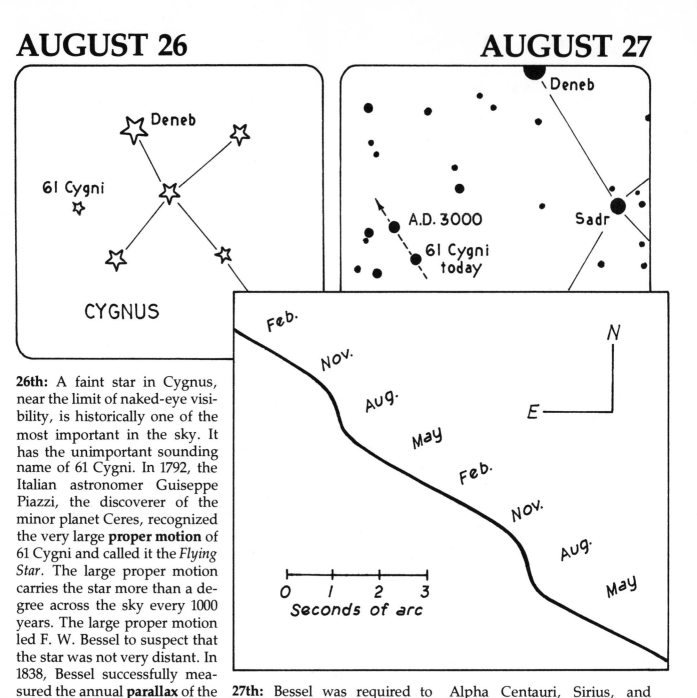

CYGNUS

26th: A faint star in Cygnus, near the limit of naked-eye visibility, is historically one of the most important in the sky. It has the unimportant sounding name of 61 Cygni. In 1792, the Italian astronomer Guiseppe Piazzi, the discoverer of the minor planet Ceres, recognized the very large **proper motion** of 61 Cygni and called it the *Flying Star*. The large proper motion carries the star more than a degree across the sky every 1000 years. The large proper motion led F. W. Bessel to suspect that the star was not very distant. In 1838, Bessel successfully measured the annual **parallax** of the star (see June 4–6), the tiny apparent change in the star's position due to the earth's motion about the sun. From the known earth-sun distance, Bessel was able to calculate the distance to the star. This was the first time the distance to any star was measured! This delicate and difficult measurement was the key that opened the universe of stars to quantitative investigation.

27th: Bessel was required to measure an angular change in the position of the star of about one ten-thousandth of a degree! In 1838 he was working at the very limit of what was possible. This is the stuff of heroic science. 61 Cygni is 11 light years away from the sun and is, we now believe, our twelfth nearest neighbor in space. It lies just outside the 10-light-year neighborhood I sketched on April 15. Of the naked-eye stars, only Alpha Centauri, Sirius, and Epsilon Eridani are closer. 61 Cygni is a binary system of two orange stars, which take something over 600 years to circle each other. A tiny unseen companion is also suspected, possibly a planet. The large proper motion of 61 Cygni, combined with our different perspective as we circle the sun, makes the star seem to snake across the sky.

AUGUST 28

AUGUST 29

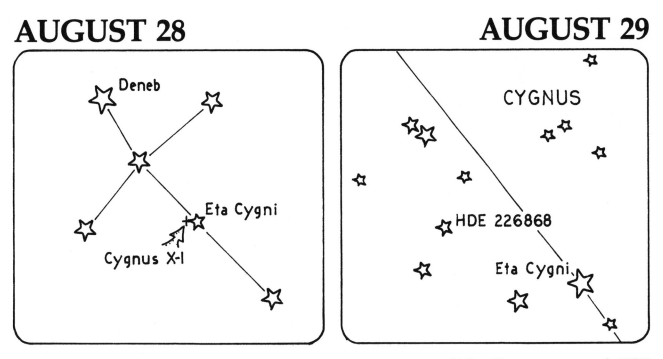

28th: In recent years, another star in Cygnus has acquired historic interest. Cygnus X-1, an object just to the northeast of Eta Cygni, is one of the strongest sources of X-ray radiation in the entire sky. Its name stands for "X-ray source 1" in Cygnus. The object was discovered in 1965 by X-ray detectors lofted high above the earth's atmosphere by rocket (the atmosphere absorbs X rays). The source was extensively studied by NASA's Uhuru X-ray telescope satellite launched in 1970. The intensity of the radiation coming from Cygnus X-1 was found to "flicker" in thousandths of a second. For something to blink as a whole at this rate, it must be very small, perhaps less than 200 miles in diameter. In 1971, radio astronomers were able to pin down the position of Cygnus X-1 more exactly. It turned out to coincide with the giant blue star HDE 226868. This large massive star could not itself be the source of X rays that flicker so rapidly. What, then, was the source?

29th: The spectrum of HDE 226868 reveals the presence of an unseen companion. The two objects rotate about one another in 5½ days. Study of the motion suggests a mass for the companion 4 or 5 times greater than that of the sun. Yet no light from the companion is visible. Many astronomers now believe that the companion of HDE 226868 is a **black hole.** As matter from the larger star is pulled off by the smaller one, it spirals down into the black hole. This accelerated material emits X rays before it disappears forever.

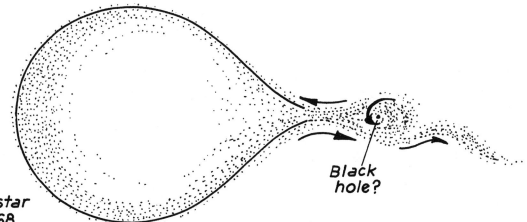

Blue giant star
HDE 226868

Black hole?

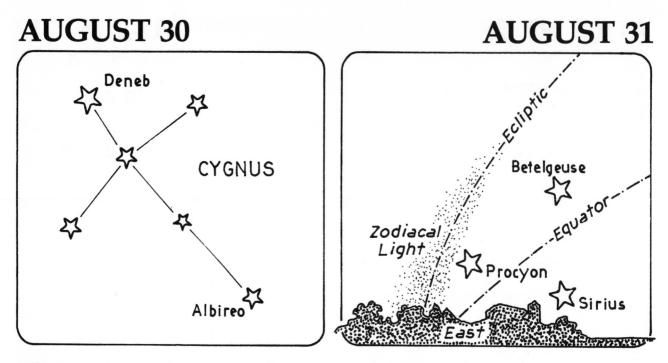

30th: Some stars, such as 61 Cygni and HDE 226868, are of interest to us because of the role they have played in the history of technical astronomy. Little Albireo (*al-BEER-ee-o*) has an esthetic interest. The name is of uncertain origin, but possibly refers to the swan's beak. In any case, it is a very special star to the observer with a small telescope. I consider it the most beautiful double star in that part of the sky accessible to northern observers. One star of the pair is blue, the other yellow. Seen in proximity, the contrast is remarkable, one of the sky's most convincing demonstrations of the paintbox variety of stars. The colors are especially prominent if the telescope is a little out of focus. The distance to the pair is about 400 light years. The two stars appear to be very far apart for a **binary star** system, and no motion has yet been observed of one star about the other. If they are linked by gravity, the bonds are very loose.

31st: If you are lucky enough to live where the sky is very dark, and if you are willing to get up early on a moonless morning, you may have a chance about this time of the year to see the *zodiacal (zoe-DYE-a-cul) light*. This is a faint glow of light, as pale as the Milky Way, which can be seen reaching up from the horizon along the ecliptic about an hour or two before sunrise. The zodiacal light is believed to be caused by sunlight reflecting from a disk of interplanetary dust that populates the inner part of the solar system. Like the planets, this dust is mostly confined to the plane of the **ecliptic.** You might also have a chance to see the zodiacal light in the evening hours toward the end of February, an hour or two after sunset. At both times of the year, the ecliptic is steeply inclined to the horizon.

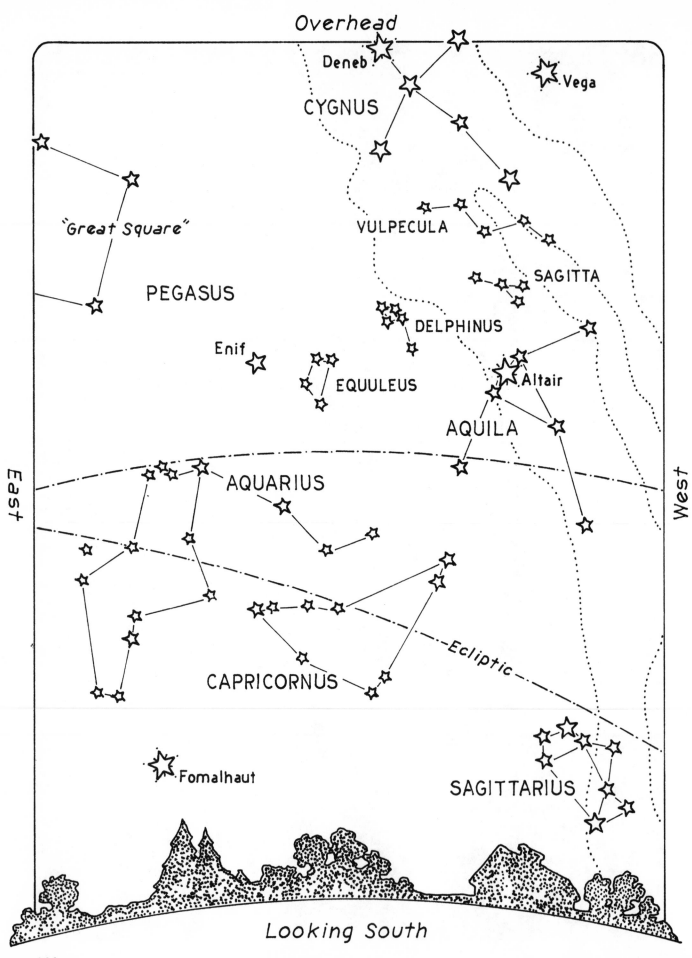

Overhead

Deneb

Vega

CYGNUS

"Great Square"

VULPECULA

PEGASUS

SAGITTA

DELPHINUS

Enif

EQUULEUS

Altair

AQUILA

East

AQUARIUS

Ecliptic

CAPRICORNUS

West

Fomalhaut

SAGITTARIUS

Looking South

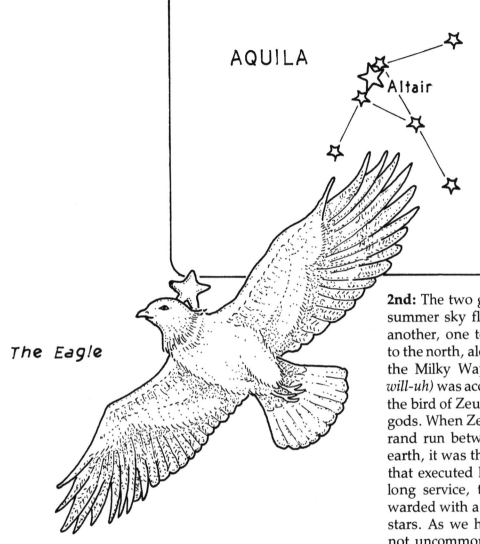

AQUILA

Altair

The Eagle

1st: There is a chill in the night air, the first hint of fall. The natural sounds of summer nights begin to fade. The three bright stars of the summer triangle move past the zenith and head for their winter setting in the west. By midnight, the winter stars can be seen creeping above the eastern horizon, Aldebaran, Rigel, and Betelguese. But in the interim, between the splendid skies of summer and the even more brilliant skies of winter, we find ourselves looking straight down out of the Milky Way Galaxy and the sky is less bright. For the next few months we shall have to sharpen our senses and learn to look for the sky's subtle wonders. During September, we shall be looking at constellations so subtle and delicate that they are usually overlooked. But first, we will survey the last of the three bright stars and constellations of summer, Altair, in Aquila the Eagle. As the swan dives south for the winter, the heartier eagle soars northward. I have represented the constellation here with the majestic American bald eagle.

2nd: The two great birds of the summer sky fly parallel to one another, one to the south, one to the north, along the stream of the Milky Way. Aquila *(ACK-will-uh)* was according to legend the bird of Zeus, the king of the gods. When Zeus needed an errand run between heaven and earth, it was the faithful Aquila that executed his wish. For his long service, the bird was rewarded with a place among the stars. As we have seen, it was not uncommon for Zeus to reward his friends with a place in the starry night. But the constellation seems to be even older than the Greek myth, and the association of these stars with the figure of a bird has been accepted among many cultures. As this starry night passes, the two birds, eagle and swan, will swing slowly across the sky toward the western horizon. Just before sunrise, Aquila will set tail first. The eagle will be followed into the "underworld" by the swan, who dives beak first into the western seas.

SEPTEMBER 3

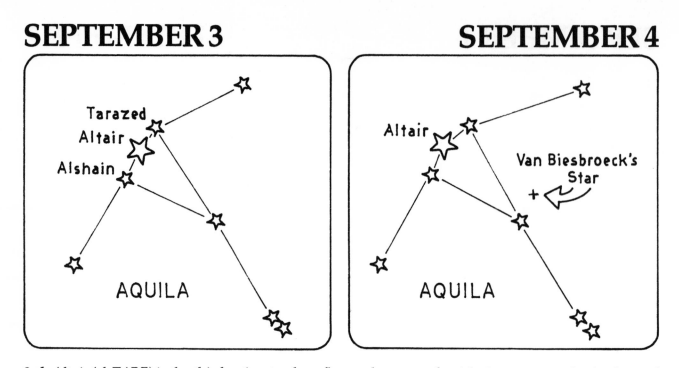

SEPTEMBER 4

3rd: Altair *(al-TARE)* is the third star of the summer triangle. It lies between Vega and Deneb in brightness. It is very close for a bright star, only 16 light years away. Altair is a white **main sequence** star, about 1½ times larger than the sun and 11 times brighter. From the spectrum of the star it can be deduced that Altair rotates extremely rapidly on its axis—once every 6 hours compared to the 25 days it takes the sun to turn. The rapid rotation tends to flatten the star and give it an ellipsoidal shape. Of course, the shape of Altair cannot be directly observed from earth. Only a few of the very largest stars have been observed as other than mere points of light. Altair means "the flying one." Alshain *(al-SHAIN)* and Tarazed *(TAR-a-zed)* also take their names from the Persian description of this constellation as a "falcon."

4th: Lost among the feathers of the eagle's back is a tiny flea of a star visible only with the very largest telescopes. This little red star has the distinction of being the least bright star known. Van Biesbroek's star is only 1/500,000 as luminous as the sun. Although it is smaller than Jupiter, its brightness, mass, and density are greater than that of the planet. The size is probably not much different than the earth. The energy source for a star this small is uncertain, since the temperature at the core may not be high enough to sustain the fusion of hydrogen to helium.

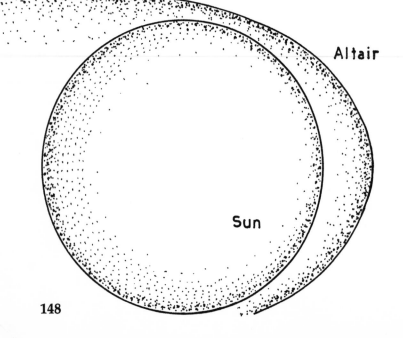

SEPTEMBER 5

SEPTEMBER 6

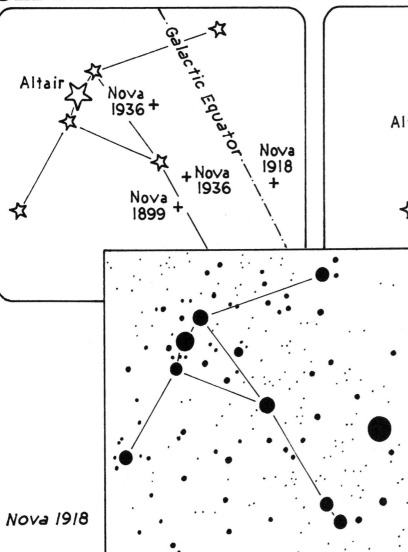

Nova 1918

5th: On June 8, 1918, a brilliant new star suddenly appeared in the evening sky. Within hours it had soared in brightness to rival Sirius as the brightest star in the sky. The **nova** was in the constellation Aquila. It was very close (as we might expect) to the galactic equator, the midline of the Milky Way. The 1918 nova was the brightest "new star" to appear in earth's sky since the great **supernova** of 1604, Kepler's Star. It has been calculated that the star which blew off its outer layers in this catastrophic way was at a distance of 1200 light years. This means that although the nova appeared in our sky in 1918, it actually occurred 1200 years earlier. Residents of that part of the universe near the star Aldebaran, opposite Aquila in the sky and 68 light years distant, are only now viewing the nova. A supernova is not merely a very bright nova. A supernova is the violent death of a massive star that scatters most of the star's material into space. A nova involves only the outer layers of a star and can reoccur.

6th: Just north of Aquila, in the dark sky between the eagle and the swan, are two little constellations that are fun to learn. If you can point them out to friends you can consider yourself a "connoisseur of constellations." The first of these is Sagitta (*sa-GIT-ta*) the Arrow. The name, of course, is related to that of Sagittarius the Archer. Sagitta is the third smallest constellation, and only Equuleus is smaller for northern observers. But the four-fourth magnitude stars of Sagitta are unmistakable on a starry night. I like to think of the constellation as an arrow that has been shot at the eagle by Sagittarius, and—fortunately—missed its target! On some ancient star maps the arrow was shown in Aquila's talons, a reminder that the eagle often served as armor bearer for Zeus. Sometimes Sagitta was identified with the arrow of Cupid; but it would be difficult to find Cupid's dart on Valentine's Day when the constellation is quickly lost in the light of the dawn sky.

SEPTEMBER 7

SEPTEMBER 8

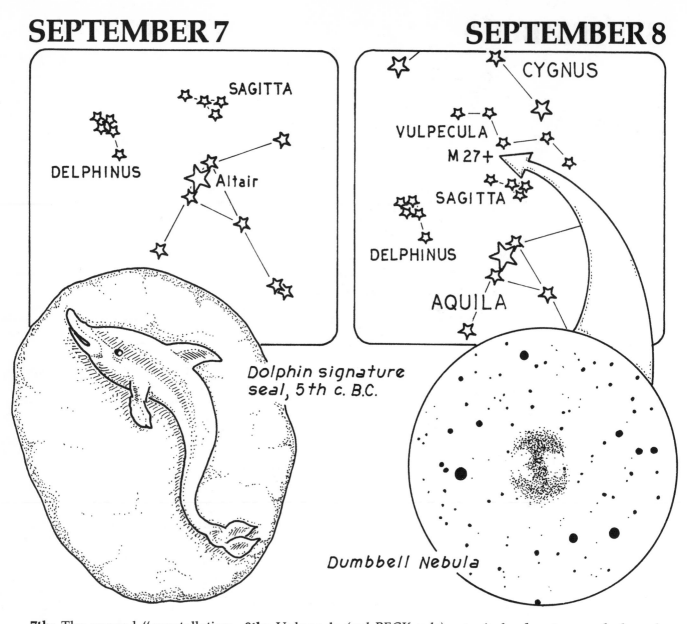

Dolphin signature seal, 5th c. B.C.

Dumbbell Nebula

7th: The second "constellation for the connoisseur" is Delphinus (del-FINE-us) the Dolphin. Only a little imagination is required to see the creature leaping gracefully out of the water, arching its back against the Milky Way. Two stars in Delphinus are named Sualocin and Rotanev. The names first appeared in a catalogue of stars compiled at the Palermo Observatory in 1814. The names were a mystery until someone noticed that *reversed* they spelled Nicholaus Venator, an astronomer at the observatory.

8th: Vulpecula (*vul-PECK-u-la*) the Little Fox is a strong contender for "least interesting constellation." Even the brightest of the stars are invisible to the naked-eye observer except on the very best nights. Vulpecula does have the dubious distinction of being last in an alphabetical list of constellations. But the little fox harbors one telescopic gem, the famous *Dumbbell Nebula,* M 27 in **Messier's catalogue.** The name derives from the shape of the nebula, which appears as two great clouds in contact. The object is a

typical **planetary nebula,** although a very near one (1000 light years). Like the Ring Nebula in Lyra, this huge bubble of gas was blown off by a star in its dying contractions. The cloud still expands outward about the dwarf star at its center. In another 50,000 years or so the nebula will have become so diffuse that it will be invisible. The star at the center will become a **white dwarf,** and finally a black dwarf. A similar demise may be the fate of our sun.

SEPTEMBER 9   SEPTEMBER 10

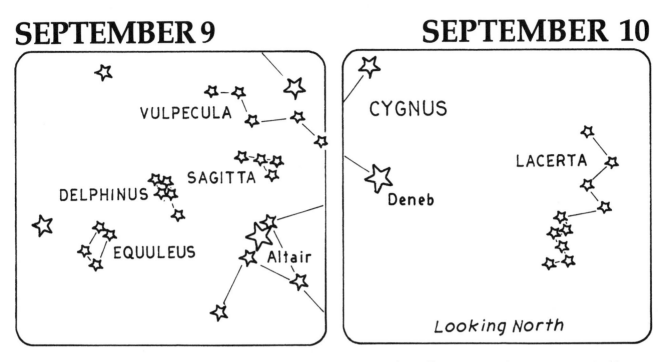

9th: Continuing our survey of the faint constellations of the summer night, we come to Equuleus *(ek-KWOO-lee-us)* the Colt or Foal, or sometimes Little Horse. Equuleus, in square degrees, is the smallest constellation in northern skies (Crux, the Southern Cross, is tiniest of all). The borders of the constellations have been officially set by the International Astronomical Union since 1930. Like all constellations, the borders of Equuleus follow north-south and east-west lines of the celestial sphere—or at least what were north-south and east-west a century ago when the borders were drawn. Because of the slow wobble of the earth's axis, the north pole of the sky slowly changes. As centuries pass, the borders of the constellations will become progressively more skewed and curved on star maps, an unhappy situation. 26,000 years from now the boundaries will again be aligned with the axis of the sky. By then, **proper motion** will have carried some stars into new constellations.

10th: Last of this series of small constellations is Lacerta *(la-SER-ta)* the Lizard. Lacerta is another of the little animals invented by Hevelius in the 17th century. To find it flicking its tail as it dashes between Cygnus and Cassiopeia, look at the star map on the next page. Little can be said for the lizard, except that it sticks its head into the Milky Way. The area is pleasant to scan with binoculars. There is another lizard, Chamaeleon, on the opposite side of the celestial sphere.

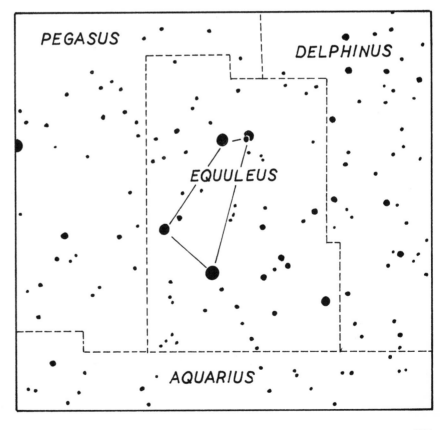

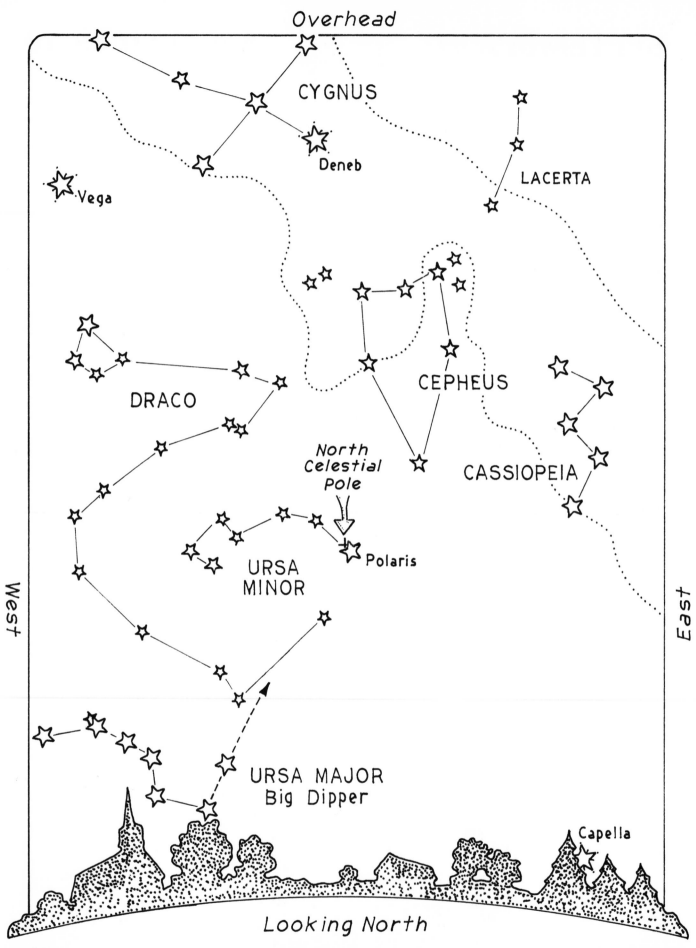

Overhead

CYGNUS

Deneb

Vega

LACERTA

DRACO

CEPHEUS

North
Celestial
Pole

CASSIOPEIA

Polaris

URSA
MINOR

West

East

URSA MAJOR
Big Dipper

Capella

Looking North

152

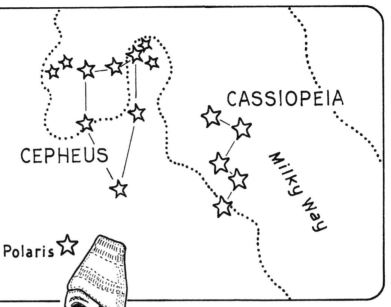

King Ashurbanipal

11th: Now let us turn for a few nights and face north, to better view constellations that are high overhead but in a more northerly direction. If the northern horizon is clear, you will see the Big Dipper skimming the treetops in the northwest. Rising in the northeast is the bright winter star Capella, preparing to replace Vega as the **zenith** star. Follow the pointers of the Big Dipper up and to the right and you will find Polaris, about halfway up from the horizon to the zenith and marking due north. Bend your head back all the way and you will see Deneb, the brilliant tail of the swan, near the zenith. To its right is the lizard (you will need a good night to see it), and directly across the pole from the dipper are the "W" stars of Cassiopeia. Between Deneb, the "W," and the pole, look for the stars of

Cepheus the King. The constellation is not particularly conspicuous; Cepheus is outshone by his consort Cassiopeia. But you should have no trouble in finding the "upside-down house" shape of the constellation.

12th: The concept of kingship may have originated in the ancient civilizations of the valleys of the Tigris and Euphrates rivers many hundreds of years before the Christian era. The identification of the stars of Cepheus with the figure of a king may be almost equally ancient. I have represented the constellation here by an Assyrian monarch of the 7th century B.C. I have placed his foot on Polaris in a posture not uncommon on old star maps. In Greek mythology, the celestial king became Cepheus, the King of Ethiopia and the husband of Cassiopeia. He figures in the story of Andromeda and Perseus that we shall tell later (see Nov. 1–2). A great loop of the Milky Way surrounds the head of Cepheus. On a moonless night when the sky is very clear you will find the king presiding from his high seat and haloed with a royal glow. As the night passes, he will lead his queen across the vault of the sky, in stately procession toward the west.

SEPTEMBER 13

SEPTEMBER 14

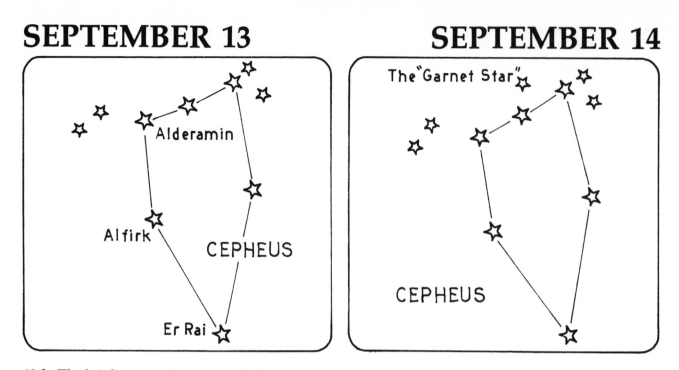

13th: The brightest star in a constellation is its *lucida.* The lucida of Cepheus is Alderamin *(al-DERR-a-min)* which means "right arm." In 5000 years, the slow wobble or **precession** of the earth's axis will bring Alderamin close to the north **celestial pole.** It was the pole star for our "cave men" ancestors of 18,000 B.C., and will be our pole star again. Alfirk *(AL-firk)* means "the flock," although the origin of the name is uncertain.

Perhaps it once referred to all of the stars of the constellation, with the exception of the star at the peak of the "house," Er Rai, whose name means "the shepherd." A shepherd and his flock evokes the memory of a simpler time when the shepherd spent nights high in the hills far from the lights of the town, with only the stars and his flock for company. What dreams then did the stars inspire?

14th: Mu Cephei, a fourth-magnitude star near the front door of the "house," is more familiarly known as the *Garnet Star.* It was given that name by the great 18th-century astronomer William Herschel. It lays claim to being the reddest naked-eye star in the sky. The Garnet Star is a huge **red giant,** not unlike Betelguese. Its size is uncertain, but if its center were where our sun is, we would be inside it.

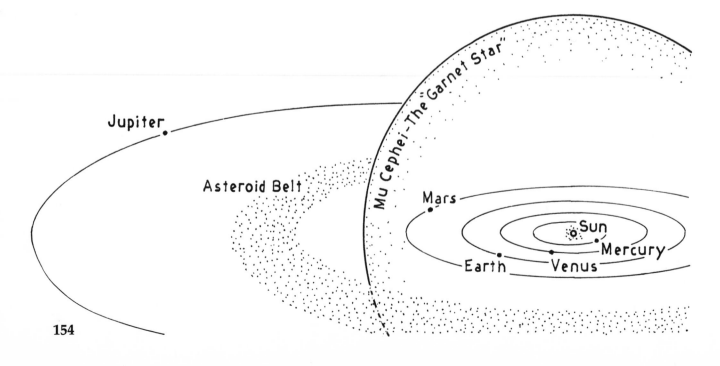

SEPTEMBER 15

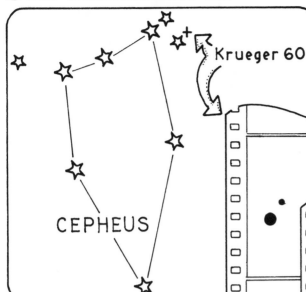

Krueger 60

CEPHEUS

15th: Very near the stellar triangle that has traditionally represented the king's head, but invisible to the naked eye, is a **binary star** remarkable for the rapidity with which the two components revolve around each other. The imaginary photographs on the filmstrip at right are "snapped" at 10-year intervals. The two stars in the system circle their common center of gravity once every 44 years. Both components are **red dwarfs,** separated by about the same distance as the sun and Saturn. (Saturn, incidentally, takes 29 years to circle the sun.) The star takes its name, Krueger 60, from its discoverer. Krueger 60 is a near neighbor of the sun. At a distance of 13 light years, it ranks as the 23rd nearest star system. It is from observations of binary systems such as Krueger 60 that astronomers are able to calculate the mass of stars. At a given separation, the speed of rotation depends upon the force of attraction. The gravitational force is proportional to the mass of the stars.

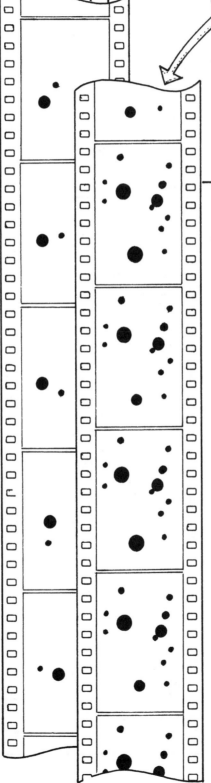

SEPTEMBER 16

Delta Cephei

CEPHEUS

16th: Delta Cephei is a near and brighter neighbor of Krueger 60. It is one of the most famous stars in the sky, at least as far as the recent history of astronomy is concerned. It is the star from which the class of **cepheid variable** stars takes its name. These remarkable stars played an important role in the discovery of the true size of our galaxy and the distances to other galaxies. The story of cepheid variables, and their role as distance indicators, will be told over the next few starry nights. The regular variations in the brightness of Delta Cephei were first noted in 1784. They are easily detected by the naked-eye observer by comparison with the surrounding stars. The imaginary photographs at left, "snapped" every 5½ days, show the rise and fall in the brightness of Delta Cephei in the context of its two bright neighbors, Epsilon and Zeta Cephei. This blink of the king's eye turned out to hold the key to some of the greatest discoveries of 20th-century astronomy.

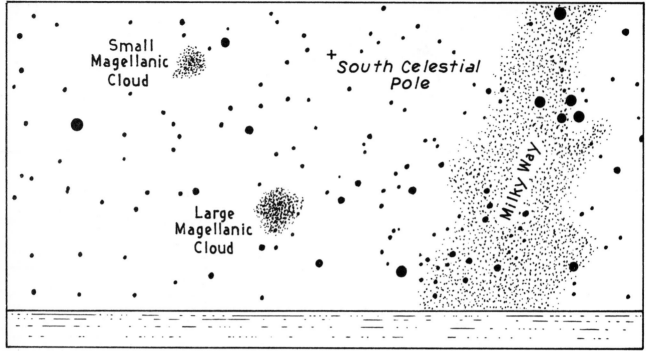

17th: Delta Cephei is a giant yellow-white star with characteristics intermediate between the massive blue stars at the top of the **main sequence** and the **red giants** such as Antares and Betelgeuse. For reasons not yet clearly understood, it undergoes regular and periodic variations in luminosity. Presumably, the star is actually swelling and contracting in size, caught in a delicate tug-of-war between gravity and outward radiation. The *period* of the variation (time between peaks) is about 5⅓ days. Delta Cephei is one of a

class of regular variable stars with similar light curves. So far, about 700 such stars have been discovered in the arms of our galaxy, with periods ranging from 1 to 100 days. A remarkable property of these **cepheid variables** was discovered in 1912 by Henrietta Leavitt, an astronomer at the Harvard College Observatory: *there is a direct relation between the luminosity of a cepheid and the period of its variation.* Leavitt found this pattern among the cepheids she identified in photographs of the Magellanic Clouds.

18th: The drawing shows the sky this evening from the latitude of Australia. The Large and Small Magellanic Clouds are prominent (see Apr. 21). The distance to the Clouds was not known at Leavitt's time, but since the **cepheid variables** she identified in each cloud *were part of a cluster,* they could be presumed to be at nearly the same distance from earth. She knew then that the *apparent brightness* of the stars was proportional to their *true luminosity.* The more luminous the star, the longer was its period of variation.

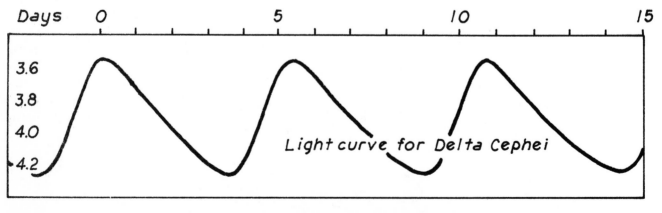

Light curve for Delta Cephei

SEPTEMBER 19

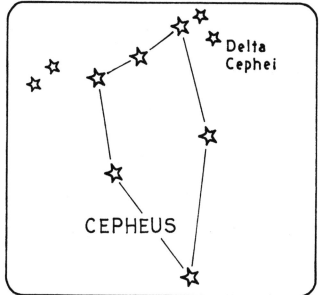

SEPTEMBER 20

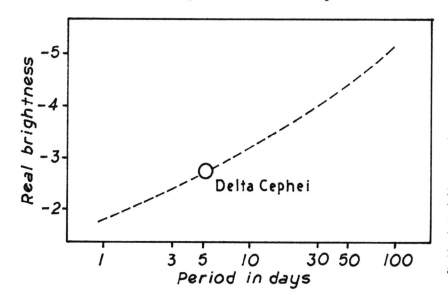

NGC 6946

19th: The distance to nearby **cepheid variables,** and therefore their true luminosity, can be determined from a statistical study of their motions. Let us assume that the cepheids in the Magellanic Clouds have the same luminosities as nearby cepheids of the same periods. Then by comparing the true luminosity of a Magellanic cepheid *as determined by its period* to the apparent brightness of the star, we can calculate the distance to the Magellanic Clouds. The distance is about 150,000 light years, which puts

the Clouds well outside our galaxy. The use of cepheids as distance indicators would be more reliable if there were one of these stars close enough to us that its distance could be directly measured by **parallax.** Unfortunately, this is not so. Delta Cephei, for example, is 1500 light years away, too far for the method of parallax to work.

20th: Harlow Shapley used the period-brightness relation of **cepheid variables** to measure the distances to globular clusters (see July 5–6). In 1923–1924, Edwin Hubble found cepheids in the star cloud we now recognize as the Andromeda Galaxy. He calculated that the Andromeda star cloud was 2 million light years away! Shapley deduced, from the distribution of the clusters, the size of our own Galaxy. Hubble proved that the Milky Way Galaxy was just one of many galactic systems. Cepheid variables can be used to measure the distance to galaxies out to about 20 million light years. NGC 6946, a beautiful spiral in Cepheus, is one of the many galaxies within this range. It is 10 million light years away.

157

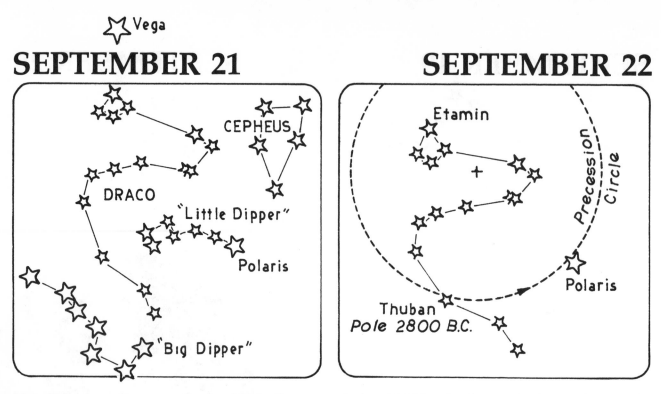

SEPTEMBER 21

SEPTEMBER 22

21st: Before we leave northern skies, let us take a look at the constellation Draco (*DRAY-ko*) the Dragon. The body of the dragon winds over much of the polar region of the sky, snaking between the Big and Little Dippers, with its head near bright Vega. Every dark sky deserves a dragon, and ours is a very ancient dragon indeed. The Babylonians believed the constellation was as ancient as creation itself, when the god Marduk slew the monster Ti'amat and created the orderly universe from the chaos of its body.

22nd: Because of the 26,000-year **wobble** of the earth's axis, the position of the poles on the celestial sphere slowly change. The poles move through the stars along circular paths, called *precession circles* (see May 22–23). Six thousand years ago the pole star was Thuban (*THEW-ban*), a star in the body of the dragon. The pole was closest to that star about 2800 B.C., within a few hundred years of the building of the Great Pyramid at Gizeh by the Egyptian pharaoh Khufu. Many people believe that the Descending Corridor which

leads into the heart of the Great Pyramid was aligned so that Thuban was visible from deep along that shaft. Exact calculation of the position of Thuban at the supposed time of construction leaves the theory somewhat in doubt. But it is clear that this and other Egyptian structures were aligned with remarkable accuracy to the celestial pole and to the rising and setting of stars. These great architectural efforts were closely tied to the birth of astronomy and mathematics.

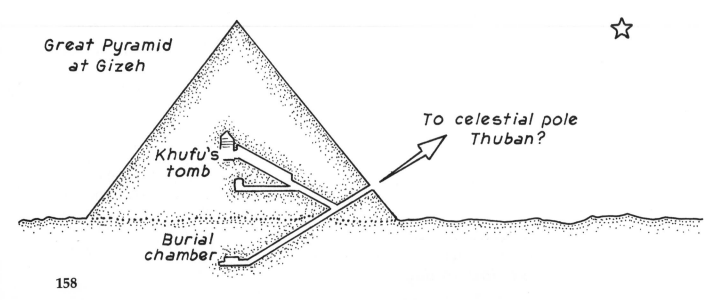

Great Pyramid at Gizeh

To celestial pole Thuban?

Khufu's tomb

Burial chamber

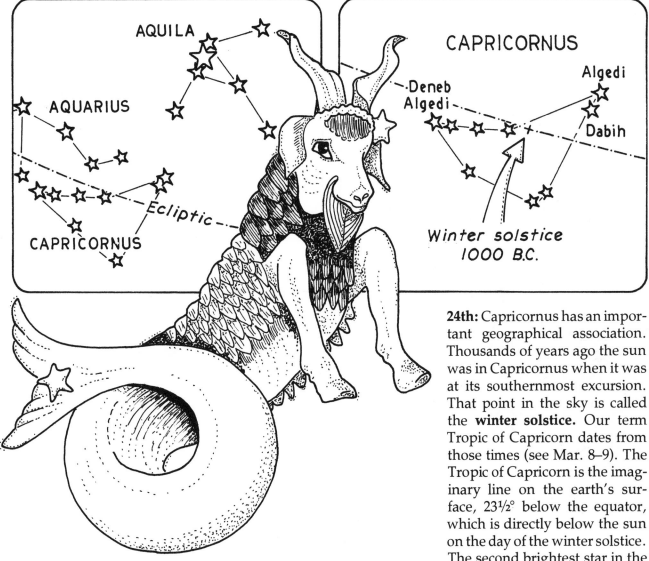

AQUILA

AQUARIUS

Ecliptic

CAPRICORNUS

CAPRICORNUS

Deneb Algedi

Algedi

Dabih

Winter solstice
1000 B.C.

23rd: Turn around again and face south. Just southeast of Aquila the Eagle are the faint stars of the constellation Capricornus. To find the constellation, it may help to refer to the larger star map at the beginning of the month. No stars in Capricornus are brighter than the third magnitude. The star nearest Aquila is Algedi *(al-GEE-dee),* which means "the goat." The brightest star in the constellation is Deneb Algedi, and you should have no trouble figuring out what that means. The traditional figure of a sea-

goat (half-goat, half-fish) goes back at least to Babylonian times. What such a preposterous monster might have represented is anyone's guess. Certainly the goat's tail is near watery constellations, and the sun in this part of the sky stood low over southern seas. My guess is that the figure of the sea-goat had something to do with the fertility of earth and sea. I have represented Capricornus by putting a fish tail on a Babylonian goat figure of the 3rd millennium B.C.

24th: Capricornus has an important geographical association. Thousands of years ago the sun was in Capricornus when it was at its southernmost excursion. That point in the sky is called the **winter solstice.** Our term Tropic of Capricorn dates from those times (see Mar. 8–9). The Tropic of Capricorn is the imaginary line on the earth's surface, 23½° below the equator, which is directly below the sun on the day of the winter solstice. The second brightest star in the constellation Capricornus is Dabih *(DAY-bee),* which means "the slaughterer." Many a goat must have been slaughtered in ancient times as sacrifice to the gods, with prayers that the weakened winter sun be restored to full strength in northern skies. With the restored sun would come the bountiful fertility of the earth. Since those faraway times, the **precession** of the earth's axis has taken the winter solstice across one whole **zodiac** constellation into the sign of Sagittarius.

SEPTEMBER 25

SEPTEMBER 26

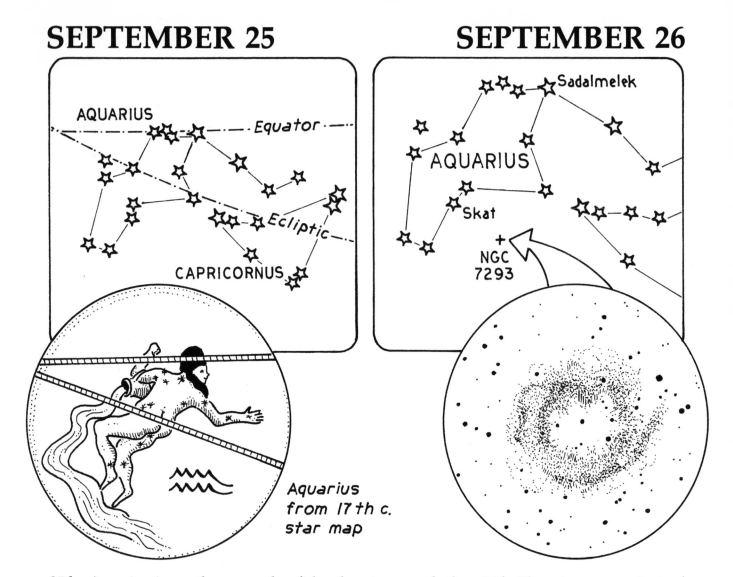

AQUARIUS

· — · — · Equator — · — · —

CAPRICORNUS

Ecliptic

Aquarius from 17th c. star map

Sadalmelek

AQUARIUS

Skat

+ NGC 7293

25th: Aquarius is another ancient constellation. The figure of the Water Carrier can be found on Babylonian artifacts. He was a celestial symbol of the significance of water in the lives of the people who lived in the river valleys of the Near East. For the Egyptians, the Water Carrier was the source of the waters of the Nile, and was especially remembered at the annual season of the flood. As far away as China the constellation has been associated with water, perhaps because the rainy season began when the sun stood in that part of the **zodiac.** Aquarius lies in close proximity to other watery constellations:

the fish, the river, and the whale. Even Capricornus has a fishtail flapping in these celestial seas. The prominence of water in this part of the sky must surely be understood in terms of the sun's journey, and the importance of water in an agricultural civilization. The brightest star in Aquarius is Sadalmelek *(sad-al-MEL-ik),* which means "lucky star of the king." Two other stars in Aquarius have names that mean "lucky." Thousands of years ago when these stars rose with the sun, winter had passed and the season of gentle rains had begun. Reason enough to feel lucky.

26th: The astronomer-priests of ancient civilizations carefully studied the sun's progress through Aquarius. Of more interest to the modern astronomer is the ring-shaped nebula in the constellation. NGC 7293 is probably the largest and the nearest of the **planetary nebulae,** those splendid bubbles of gas blasted from the surfaces of stars at the end of their lives. The distance to NGC 7293 is about 100 light years, and the object fills an area of the sky half as wide as the moon. If you own a good pair of binoculars and wait for a perfectly dark night, you might catch a hazy glimpse of this beautiful object.

160

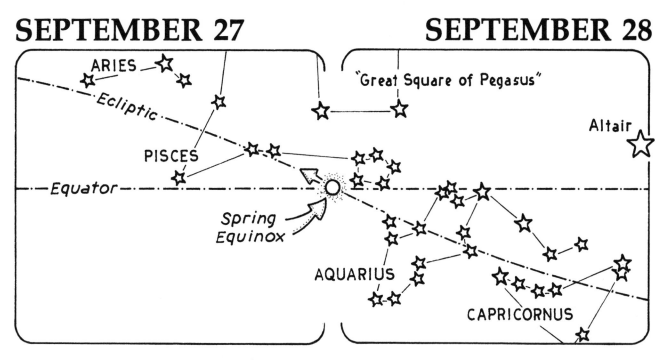

27th: Before we leave the constellation Aquarius we should refer to the "Age of Aquarius" heralded in the song from the musical *Hair* and books on astrology. The spring season begins as the sun, moving along the **ecliptic,** crosses the **celestial equator** into northern skies. That crossover point in the sky and the moment the sun is there are called the **spring (or vernal) equinox.** Because of the 26,000-year **wobble** of the earth's axis, the equator slowly tilts and shifts with respect to the ecliptic. The position of the crossover point creeps along the ecliptic against the background of the constellations. This slow change is called the *precession of the equinoxes.* Thousands of years ago, when astrology was born, the sun was in Aries as it crossed the equator. The equinox is presently in the constellation Pisces. It continues to slide toward Aquarius at a rate of about 1½° per century. In another 600 years it will cross the border into that constellation.

28th: Astrology is the belief that stars influence earthly affairs and that the **zodiac** exerts a particular influence on our lives. The "sign" of the **spring equinox** is an important controlling factor. Astrologers believe we are already feeling the effects of "the dawning of the Age of Aquarius." The new age will be marked (according to believ-ers) by a harmonious blending of the sciences and the humanities. Art and science will flourish together and humankind will live at peace with one another. Astronomers give little credence to the archaic superstitions of astrology, but let us hope that such a vision comes to pass—with or without the help of the stars.

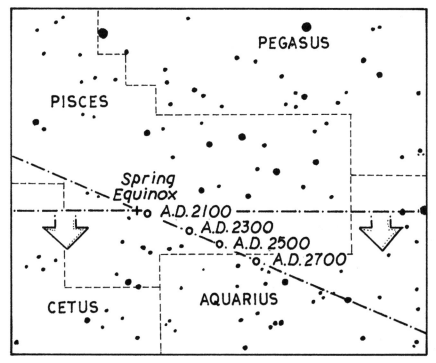

SEPTEMBER 29

SEPTEMBER 30

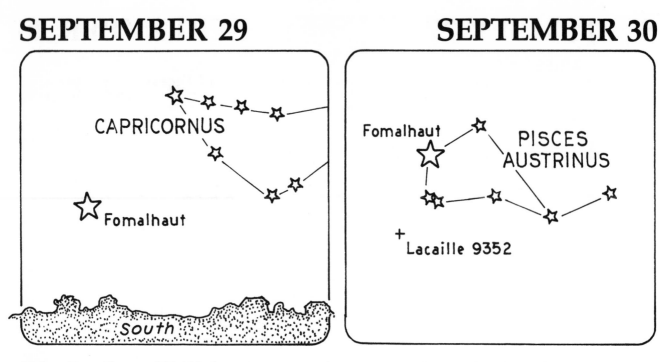

29th: Fomalhaut *(FOAM-al-ought)* gets far less attention than the star deserves. For most observers at mid-northern latitudes, it is the southernmost first-magnitude star. During the late summer and autumn Fomalhaut dominates the southern sky, standing only a handspan or so above the horizon. The star is often called "the solitary one." There are no other prominent stars in this part of the sky (see Oct. 3–4).

The drawing below shows the southernmost first-magnitude stars that can be seen from the latitudes of major U.S. cities. These are mathematically exact horizons; the effects of local geography, buildings or vegetation, city lights, and the atmosphere will all limit your view. Fomalhaut means "mouth of the fish." It is in the constellation Pisces Austrinus, the Southern Fish.

30th: Also in Pisces Austrinus, but invisible to the naked eye, is the star Lacaille 9352. This is the 17th nearest star system to our own. It is 12 light years away. Lacaille 9352 is another of those countless little red dwarfs that clutter the galaxy. It was the large **proper motion** of the star, 1° across the sky every 500 years, that attracted attention to the fact that it was a near neighbor.

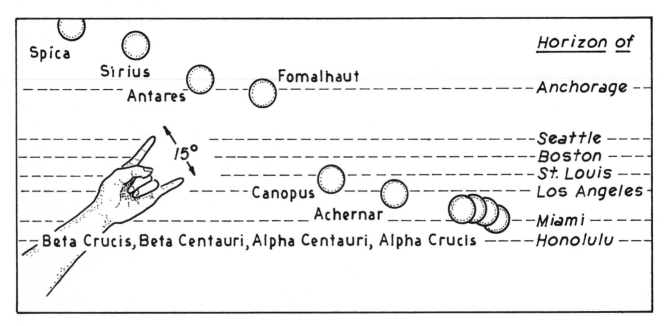

Overhead

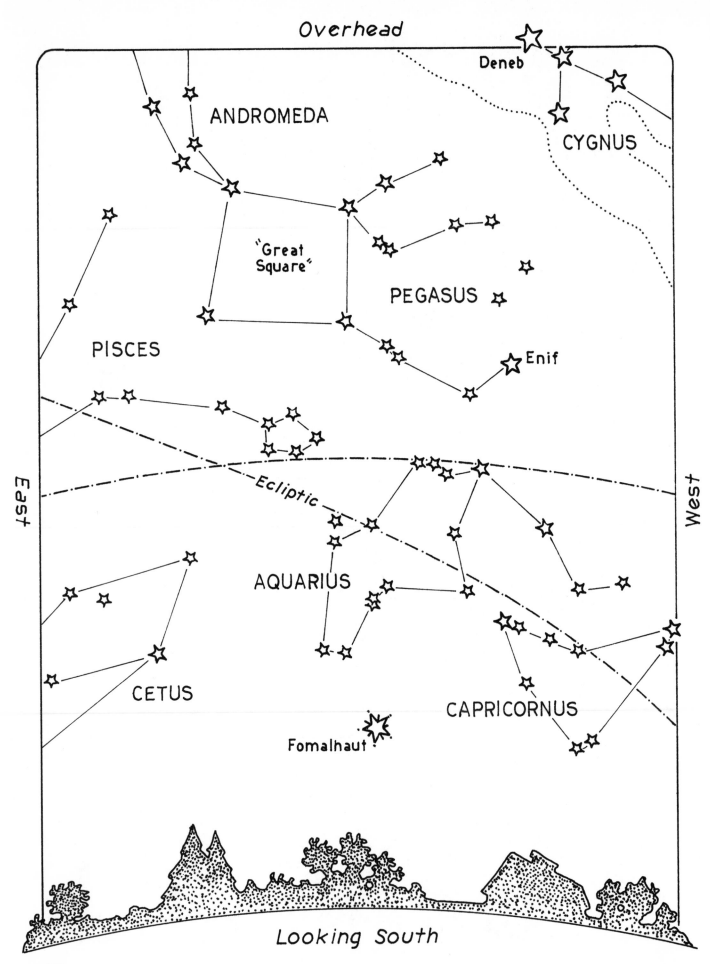

Deneb

ANDROMEDA

CYGNUS

"Great
Square"

PEGASUS

PISCES

Enif

East

West

Ecliptic

AQUARIUS

CETUS

CAPRICORNUS

Fomalhaut

Looking South

OCTOBER 1

OCTOBER 2

1st: According to the National Climatic Center, October is the least cloudy month of the year for much of the eastern and southern parts of the United States. From Maine to Texas you have at least a 50 percent chance of having a starry night. If you live in the western United States, July is probably your least cloudy month of the year. In the Great Plains states and Michigan, your best bet for starry nights may be August. But although October is relatively cloud-free for the greater part of the population of the United States, the evening sky in October presents one of the least interesting expanses of the entire celestial sphere. A glance at the star map at the left will show you what I mean. Only one first-magnitude star, Fomalhaut, appears on the map. And that solitary star of southern skies is probably the least familiar of all first-magnitude stars. But nature has

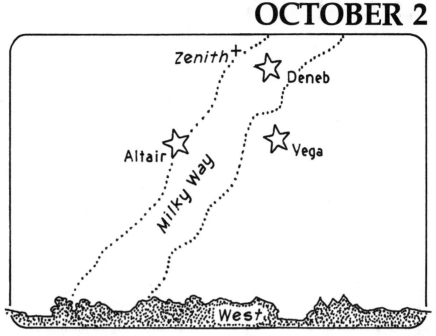

a way of spreading her gifts equally across the year, and what is missing in the night skies of spring and fall is compensated for by quickened pleasures on the earth. Now is the season of mushrooms, the migration of birds, and the colorful transformation of the deciduous forests.

2nd: Fomalhaut is likely to be lost in southern haze. And so the sky this evening continues to be dominated by "summer" stars, the triangle of Deneb, Vega, and Altair, still high but sinking fast toward the west. It also remains a good time to look for "shooting stars," but don't expect to see as many as you saw in July or August. October has one important meteor shower of its own, the *Orionids*. These meteors are associated with the orbit of Halley's Comet, and remind us annually of the periodic return of that long-tailed visitor to our starry nights (next visit in 1986). The "shooting stars" of this shower will appear to radiate from the place where Orion hides over the northeastern horizon. Also in the northeast you will see Capella rising, a prelude to the brilliant stars of winter.

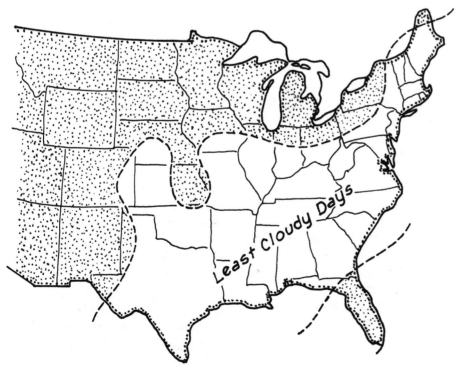

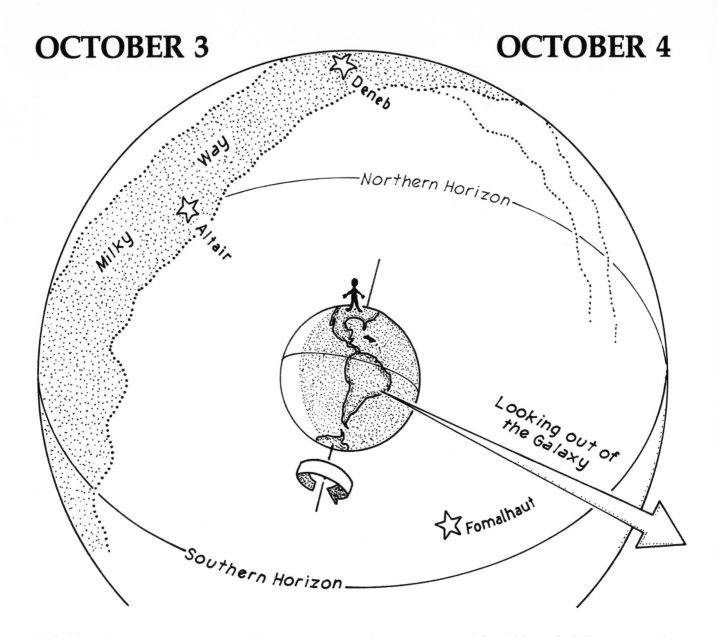

3rd: There is a reason why that part of the sky we see toward the south this evening is relatively devoid of bright stars. When we look toward Fomalhaut, we are looking directly out of the flat spiral of the Milky Way Galaxy. Since the Galaxy is only a few thousand light years thick at the position of the sun, we are only 1000 light years away from the vast emptiness of intergalactic space. Travel more than 1000 light years in the direction of Fomalhaut and you will leave the stars of our galaxy behind.

In your rearview mirror, the sun would fade to invisibility, lost among the myriad stars of the Milky Way. Fomalhaut, the "solitary one," is not a distant star. It is only 23 light years away: of the first-magnitude stars, only Sirius, Procyon, and Altair are closer. Most of the stars you would pass on your journey out of the galaxy would be sun-sized or smaller. Beyond Fomalhaut there are no giants like Rigel or Deneb to add their brilliance to the starry night.

4th: Although I have no evidence to support me, I like to believe that the relative emptiness of this part of the sky was one of the reasons for the prevalence of "watery" constellations. There are few "landmarks" in the southern October sky, just as there were few landmarks available to those Mesopotamian sailors who left (by design or chance) the safety of the rivers Tigris and Euphrates to venture upon southern seas. Sea and sky met on the southern horizon—and in the imagination of the first astronomers.

166

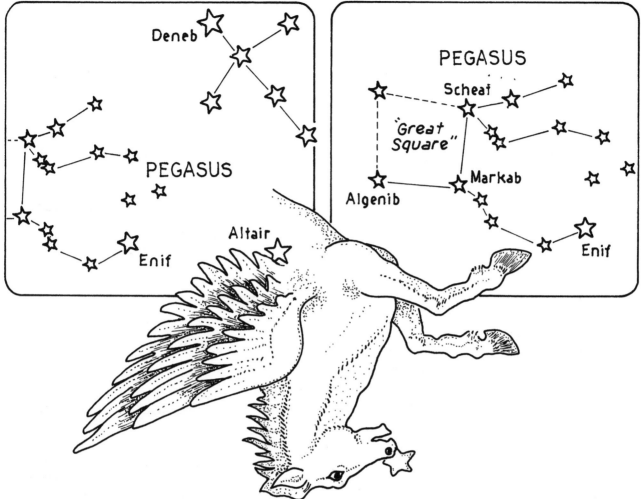

5th: The huge constellation Pegasus the Winged Horse is one of the most impressive features of October skies. According to legend, Pegasus was fashioned by the sea god Poseidon from beach sand, sea foam, and blood that fell from the severed head of the snaky-haired Medusa when she was slain by Perseus. Few inventions of Greek mythology are as attractive as this magnificent flying steed. A later embellishment of the myth associates Pegasus with the story of Andromeda, which I shall tell next month. Like Taurus, Pegasus has been traditionally shown with an incomplete body.

Perhaps parts of both constellations were removed by the Babylonians to create the new constellation Aries, when the **spring equinox** moved into that region of the sky. Or perhaps the figures of the bull and the winged horse were imagined to be rising from the "watery" part of the sky near Pegasus. The brighter stars of Pegasus take their names from parts of the horse. Enif (*EN-if*) means "the nose." Markab (*MAR-kab*) means "saddle," though Pegasus is seldom shown saddled. Scheat (*SHE-at*) means "upper leg," and Algenib (*al-GEE-nib*) refers to "the wing" or possibly "the side."

6th: The most easily recognized feature of the constellation is the so-called *Great Square of Pegasus.* This almost perfectly square configuration of four stars stands high between summer and winter constellations. The star at the northeast corner of the square actually belongs to the constellation Andromeda, but only barely. The square is 15° on a side, or a handspan at arm's length. It is oriented almost exactly north-south, east-west. As we shall see, the Great Square is a great help in visualizing the celestial grid which astronomers have imposed on the sky.

167

OCTOBER 7

OCTOBER 8

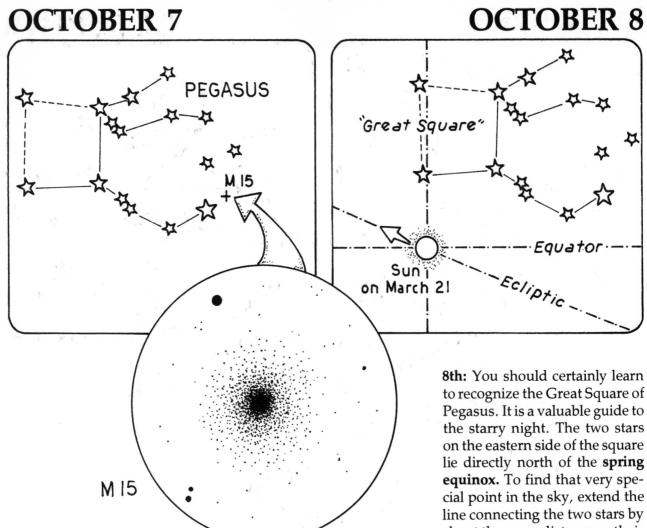

7th: Not far from the nose of the Flying Horse, like a puff of misty breath from his flared nostrils, is the beautiful **globular cluster** M 15. This is an object for the amateur telescope. It is one of the finest globular clusters in northern skies, but cannot compare to the magnificent M 13 in Hercules. There are hundreds of thousands of stars in this ball-shaped miniuniverse. It lies about 35,000 light years away and is part of the great halo of globular clusters that surrounds the Milky Way Galaxy and presumably other spiral galaxies. The distance to M 15 was measured using the method of **cepheid variables** which was discussed earlier (Sept. 17–20).

8th: You should certainly learn to recognize the Great Square of Pegasus. It is a valuable guide to the starry night. The two stars on the eastern side of the square lie directly north of the **spring equinox.** To find that very special point in the sky, extend the line connecting the two stars by about the same distance as their separation. The place is empty; unlike the north **celestial pole,** we are not lucky enough to have a star to mark the spot. The word *equinox* ("equal night") refers to the fact that when the sun stands on the celestial equator, day and night are of equal length.

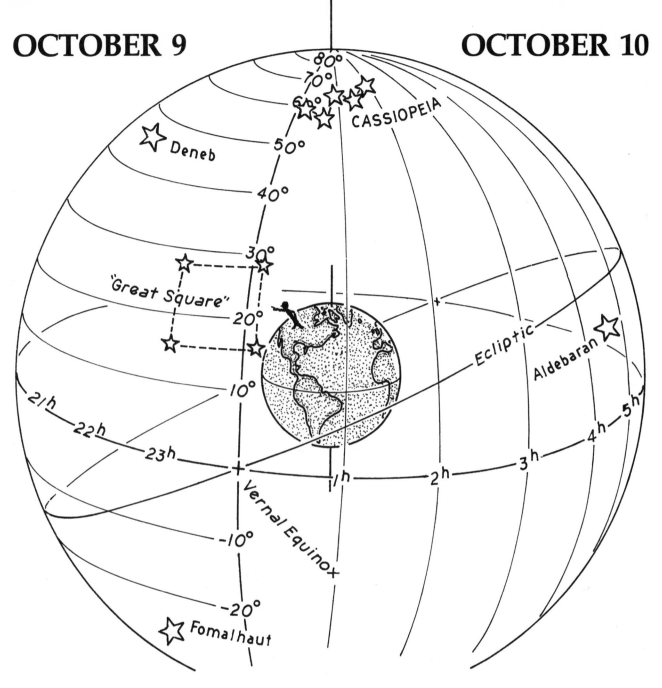

9th: You are familiar with the use of latitude and longitude to specify locations on earth. A very similar system is used on the imaginary sphere of the sky. Like terrestrial coordinates, celestial coordinates are based on the axis of the earth's rotation. The earth's poles projected onto the celestial sphere become the **celestial poles.** The earth's equator, projected onto the sky, becomes the **celestial equator.** The positions of stars north and south of the equator are designated like places on earth, except the coordinate is called **declination** rather than latitude. The declination of a star on the celestial equator (Mintaka in Orion's belt) is 0° (zero degrees). The declination of Polaris, which stands near the north pole, is +90°. Deneb, halfway from equator to pole, has declination +45°. Fomalhaut has declination 30° south of the equator, or −30°.

10th: The east-west coordinate in the sky is called **right ascension.** Right ascension is measured in hours, minutes and seconds, like time. The sky's equator is divided into 24 hours (the time it takes the earth to rotate under the sky), eastward from the **spring equinox.** The right ascension of the stars on the east side of the Great Square, and of Beta Cassiopeia, are close to zero hours. Aldebaran has right ascension 4 hours, 33 minutes.

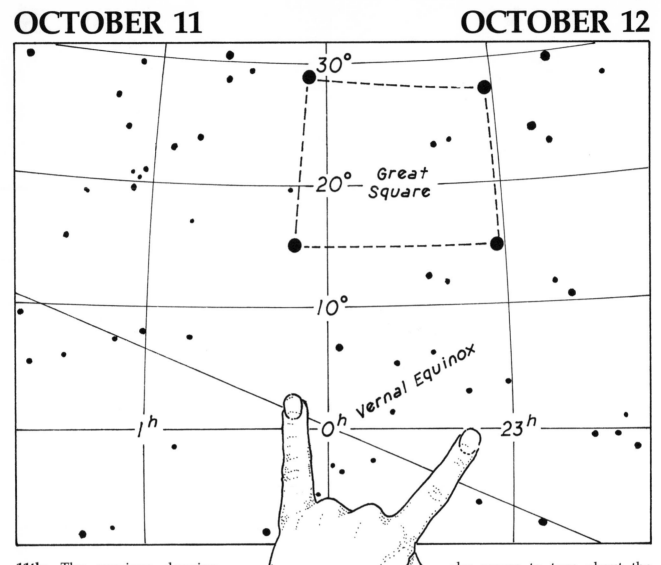

11th: The previous drawing showed the imaginary celestial sphere from "outside." Here we are looking from inside. The Great Square offers an excellent chance to imagine lines of **right ascension** and **declination** against the background of stars. The celestial equator arches from due east on the horizon to due west, passing just below the Great Square. The angle between your spread fingers, hand at arm's length, is about 1 hour of right ascension along the equator, or 15° of declination north-south. The Great Square is 1 hour wide east-west and 15° high north-south.

12th: With a little practice you should be able to imagine the whole spider's web of celestial coordinates projected onto the sky. These imaginary lines of **right ascension** and **declination** are fixed on the sphere of stars, just as lines of longitude and latitude are fixed on the earth's surface. As the earth turns eastward under the stars, the sky seems to turn about the earth. The stars *and* our imaginary network of lines turn toward the west, moving one handspan (at the equator) each hour. Line up the western edge of the Great Square with some reference object (a chimney or tree branch). Look again an hour later. Now the eastern edge of the square will be aligned with your reference. As the night passes, the hour lines of right ascension pass overhead, like the turning face of a great 24-hour clock. The choice of the **vernal equinox,** by the way, to mark zero hour on the "clock" is entirely arbitrary.

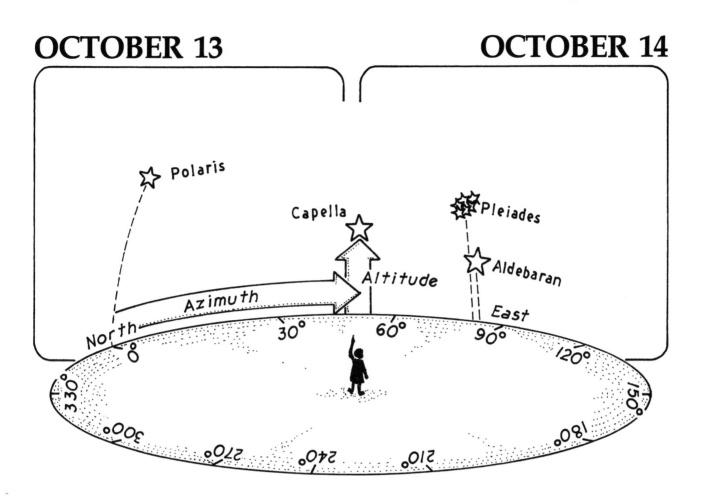

13th: The drawing above shows the northeastern sky tonight at about 10 o'clock. Capella, the Pleiades, and Aldebaran are rising. By midnight, Orion too will be above the horizon. One way observers can describe the position of a star in their sky at a certain time is by the use of horizon coordinates, azimuth and altitude. The **azimuth** of a star (or any other object in the starry night) is the angle measured eastward along the horizon from the north point to the place on the horizon directly below the star. For example, in the drawing above, the azimuth of Capella is 50°. The azimuth of Polaris is 0°, and Aldebaran and the Pleiades have an azimuth of about 80°.

14th: The **altitude** of a celestial object is the angle measured upward from the horizon to the object (the ideal horizon, as if you were at sea). A point that is on your horizon has an altitude of zero degrees. Your *zenith*, the point in the sky directly over your head, has altitude 90°. Your *nadir*, the point on the unseen sky directly beneath your feet, has altitude −90°. In the drawing above, Capella has an altitude of about 34°. Polaris, the North Star, has an azimuth of 0° for all observers, but the altitude of Polaris will depend upon the observer's latitude on earth. If you were standing at the north pole, Polaris would stand at your zenith, altitude 90°. If you lived on the equator, Polaris would lie on the northern horizon, altitude 0°. The terrestrial latitude of the observer in the drawing is 40° north; accordingly, Polaris has an altitude of 40° above the horizon. The system of **equatorial coordinates** discussed on October 8–12 is fixed *in the sky*. The **right ascension** and **declination** of a star do not depend on the latitude of the observer or the time of day. You will note, however, that azimuth and altitude are fixed *in the horizon* of a particular observer. The **horizon coordinates** of stars change as the sky turns (except, approximately, for Polaris).

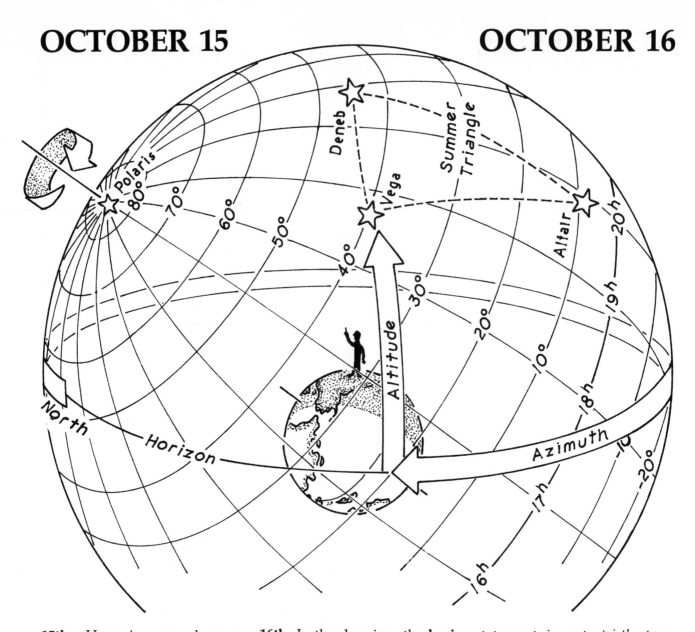

15th: Here is our observer again, tonight. This time we are looking at him from outside the imaginary sphere of the sky. The time is toward the middle of the evening. The familiar *Summer Triangle* is moving down toward the northwestern horizon. The **equatorial coordinates** of Vega are 18 hours, 36 minutes right ascension and +39° declination. This is usually written in the abbreviated fashion R.A. 18h 36m, dec. +39°. Altair has coordinates R.A. 19h 50m, dec. +9°.

16th: In the drawing, the **horizon coordinates** of Vega are approximately 290° azimuth and 35° altitude. As the night passes and the sky turns, the horizon coordinates of Vega will change. The star moves toward the northwestern horizon; its azimuth will increase and its altitude decrease. Vega will set shortly after midnight, at a point on the horizon near the letter "O" in the drawing (azimuth 320°, altitude 0°). Meanwhile, the **equatorial coordinates** of Vega do not change. Actually, the last

statement is not *strictly* true. The right ascension and declination of nearby stars show tiny periodic changes called **parallax,** reflecting the annual motion of the earth. The **proper motion** of stars adds its effect, and is also greater for nearby stars. Finally, the 26,000-year **wobble** of the earth's axis causes celestial coordinates to change significantly over long periods of time. (Most present star maps are based on 1950 coordinates. New maps will show the positions of stars in the year 2000.)

OCTOBER 17   # OCTOBER 18

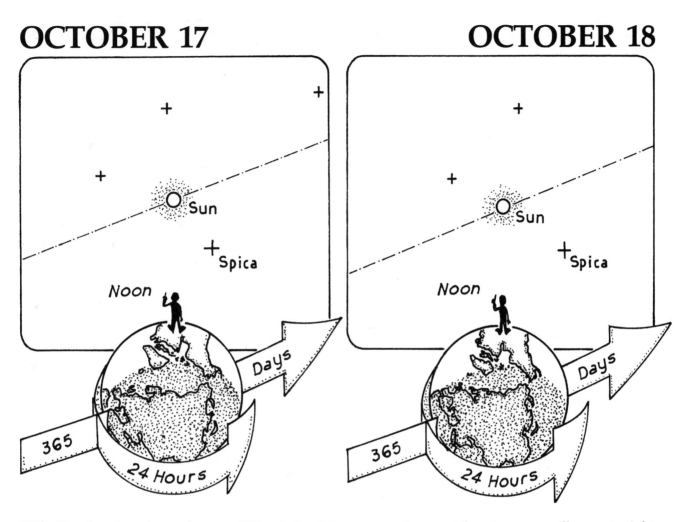

17th: The drawing above shows the sun at noon today. For our North American observer, the sun is in the south, about halfway up from the horizon to the **zenith.** Our observer cannot see the stars in that part of the sky because of the overpowering light of our nearby star scattered through the earth's atmosphere. But I have indicated the position of a few background stars with crosses. The sun is in the constellation Virgo, only about 2° from brilliant Spica. The sun covers ½° of the sky, or about half the width of your little finger held at arm's length. I have drawn it to scale. The dashed line is the **ecliptic,** the path of the sun's apparent motion across the celestial sphere.

18th: A day later, a new drawing. Twenty-four hours have passed on the clock. The earth has turned once on its axis, carrying our observer around through the starry night. Now it is noon again, and the sun is again due south, halfway between eastern and western horizons. But something has changed. We have moved 1/365 of our great annual journey around the sun. As viewed from the earth, the sun has moved almost 1° against the background stars. Of course, this is an apparent motion, not a real one. The change is slight, but significant. In the course of the year this degree-by-degree motion will carry the sun all the way around the celestial sphere.

The stars we will see at night opposite the sun will slowly shift. Six months from now Spica will be shining brightly at midnight. Our clocks are set to stay with the sun, and therefore keep *solar time.* By our clocks, Spica is today at the same point in the sky 4 minutes earlier than yesterday. If we adjusted the rate of running of our clocks so that 24 hours had passed when the stars (say, Spica) returned to the same place in the sky, the clocks would be keeping star time or *sidereal (sigh-DEER-ee-al) time.* Astronomers use sidereal clocks to set their telescopes for looking at the stars, but they use a solar clock to decide when it is time for their midnight snack.

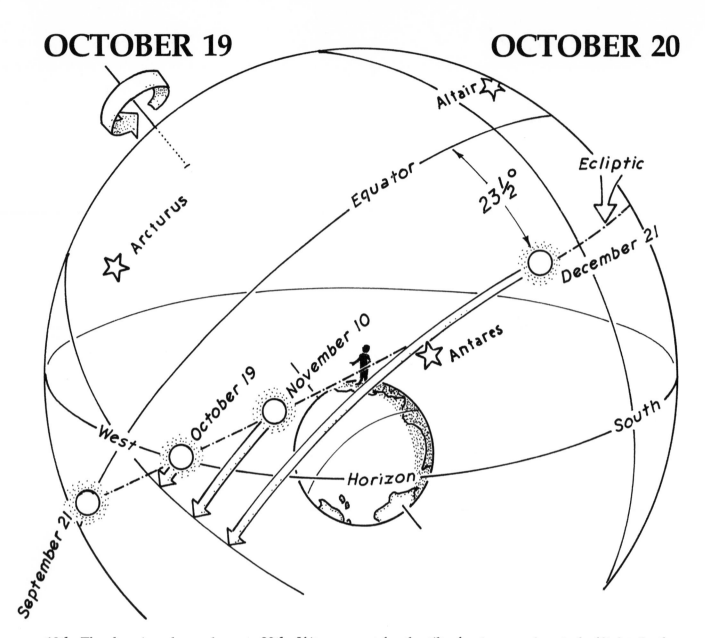

19th: The drawing above shows the sky tonight just at sunset. The **autumnal equinox** is past. The sun has moved below the equator in its journey along the **ecliptic.** It sets a little to the south of west. It also sets a little earlier each night, and rises a little later each morning. If the earth's axis were not tipped to the plane of its orbit, the sun would always be on the sky's equator. It would always rise due east and set due west, and days and nights would always be exactly 12 hours long.

20th: If it were not for the tilt of the earth's axis we would not have seasons. Now, as winter approaches, the sun spends less and less time above the horizon, at least for observers in the northern hemisphere. The days get shorter and the starry nights longer. The time from sunset to sunrise tonight will be a little more than 13 hours (at a latitude of 40°N), and the interval between sunrise and sunset tomorrow, a little less than 11 hours. Of course, morning and evening twilight, an effect of the earth's atmosphere, will extend the actual period of light. By the day of the **winter solstice,** December 21, the number of hours between sunset and sunrise for mid-latitude observers will increase to nearly 15. Long nights for stargazers—but cold ones! Two things cause the colder weather. As we have seen, the sun now spends less time above the horizon. But more significant is the steeper slant at which sunlight strikes the earth's surface in the northern hemisphere, spreading more thinly the heating effect of the sun's radiation.

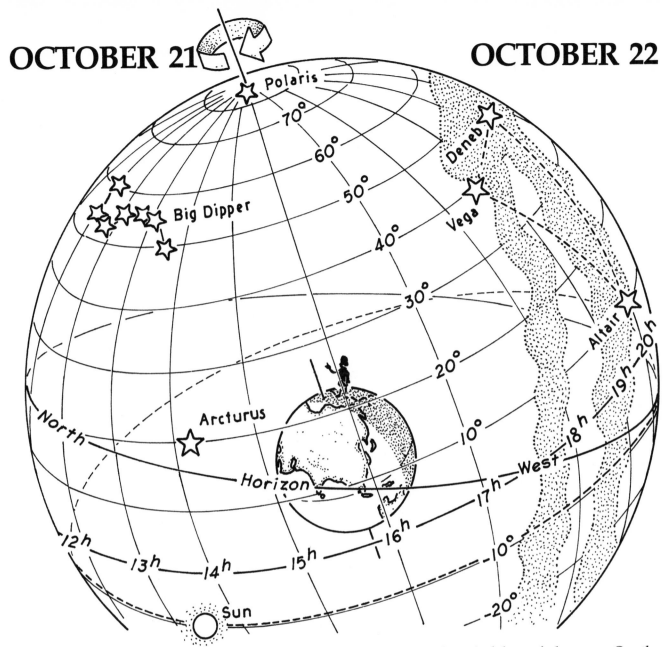

21st: The sky we see this starry night also depends on where we live on the earth's surface. Most of the maps and drawings in this book have been prepared for a hypothetical observer at a latitude of about 40°N. But the drawing above shows an observer at Prudhoe Bay, Alaska, above the arctic circle. It is evening and Polaris is high overhead, near the **zenith.** The Big Dipper makes a high circle around the top of the sky. Arcturus has already set in St. Louis, but for our Alaskan observer the star dips to skim the Arctic Sea.

22nd: The most striking celestial phenomenon for observers north of the arctic circle is the *midnight sun.* You will see in the drawing the path the sun takes as the sky turns today. At Prudhoe Bay in Alaska the sun will rise only briefly above the horizon and stay very low in the southern sky. In December, when the sun is 20° or more below the equator, it will *never* rise above the horizon for our Alaskan observer! She will experience 24 hours of darkness with only the rosy glow of a dawn that never comes to remind her of the sun. On the other hand, in June the sun is further north of the celestial equator than the star Arcturus. Then, like Arcturus tonight, the sun will never set. At "midnight" our observer will see the sun standing low on the northern horizon. Twenty-four hours of daylight! The effect is reversed in the southern hemisphere. In December an antarctic observer has 24 hours of daylight. The *arctic* and the *antarctic circles* define the limiting latitudes for the phenomenon of the "midnight sun."

175

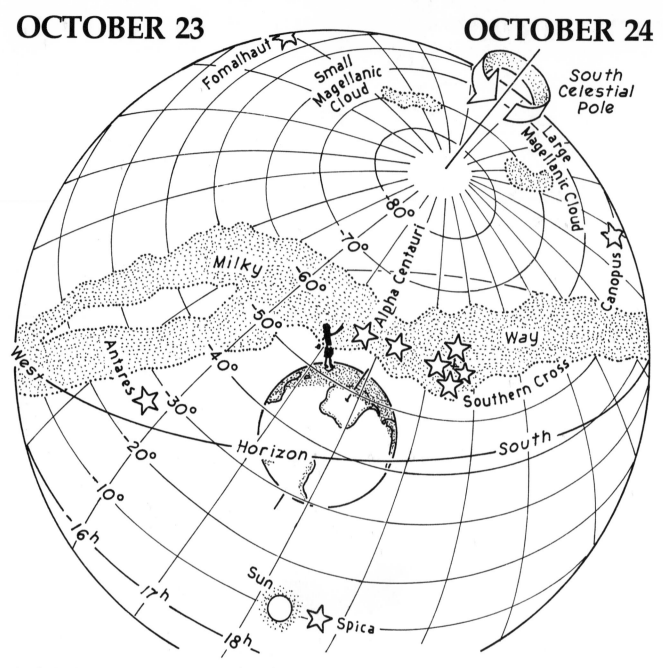

23rd: Let's take one more trip in the imagination, this time to New Zealand. The drawing shows the sky this evening for an observer in those southern islands. Polaris is forever hidden beneath the earth. Deneb, which is close to our **zenith,** is on the northern horizon for the New Zealander. Fomalhaut, which pretty much marks the southern limit of stars available to our typical northern observer, stands near the zenith.

24th: As our starry nights get chillier, the New Zealander feels a growing warmth as the sun continues its steady journey southward along the **ecliptic.** There is no star to mark the southern pole, only a dark and empty gulf of sky. But the stargazer in New Zealand has wonders to observe that we have never seen. The brilliant stars Alpha and Beta Centauri, Canopus, and the stars of the Southern Cross are in the night

sky all year round, since they are so close to the southern pole that they never set. The Milky Way is especially bright in this part of the sky, and is accompanied by the two hazy Clouds of Magellan. Alpha and Beta Centauri, by the way, have proper names. Alpha Centauri is also Rigel Kentaurus, which means "foot of the centaur" (you will recall Orion's foot). Beta is Hadar, which means "settled land."

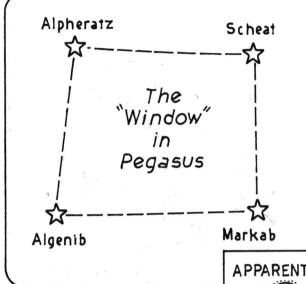

The "Window" in Pegasus

Alpheratz · Scheat · Algenib · Markab

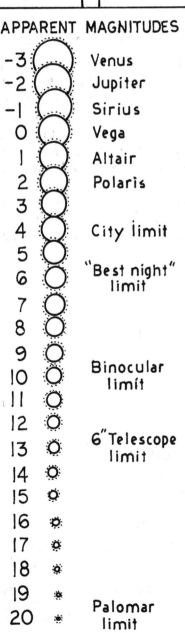

APPARENT MAGNITUDES

Magnitude	Object / Limit
−3	Venus
−2	Jupiter
−1	Sirius
0	Vega
1	Altair
2	Polaris
3	
4	City limit
5	
6	"Best night" limit
7	
8	
9	
10	Binocular limit
11	
12	
13	6" Telescope limit
14	
15	
16	
17	
18	
19	
20	Palomar limit

25th: For the past few weeks of starry nights, we have been imagining that all of the stars reside on a single sphere which encloses the earth. We called this useful fiction the **celestial sphere.** In fact, this is how people believed the sky to be until the time of Galileo and Newton in the 17th century. We now realize, of course, that celestial objects are distributed in a vast (infinite?) three-dimensional space that stretches beyond the limit of our most powerful telescope. And the earth, we know, is in no way central to that space. Our Galaxy, our star, our planet, perhaps we ourselves are utterly typical. The view from here is in no way special. During the next week of starry nights, we shall use the Great Square of Pegasus as a "window" for peering deeper into the space of the universe. And incidentally, because of the finite velocity of light, we shall be looking deeper and deeper into the past.

26th: On a typical city night the "window" of Pegasus might be completely empty, with only the 2nd-magnitude stars at the corners visible to the eye. These four stars all lie between 100 and 500 light years away. They are relatively bright because they are blue and red giant stars. If you are far from city lights and the night is clear, several other stars may appear inside the window. Under absolutely ideal conditions on a moonless night you may see a dozen or more stars. All of these stars are neighbors in our corner of the Milky Way Galaxy. They are typically tens or hundreds of light years distant. Recall that as we look toward Pegasus, we are looking out of the flat disk of the Galaxy. You will also recall that, at the position of the sun, the Galaxy is only several thousand light years thick. In this naked-eye view through the "window" we are seeing a significant part of the distance to the edge of the Galaxy.

27th: With a good starry night and a good pair of binoculars you might see hundreds of stars through the Pegasus "window." They will all be stars in a pyramid-shaped wedge extending right out to the edge of the flat disk of the Galaxy. The drawing below shows a cross section of the Milky Way Galaxy in the region of the sun.

28th: A telescope reveals even more stars through our "window." Stars in the wedge not bright enough to be seen with binoculars now come into view. An 8-inch telescope, for example, will show the little group of 11th- to 14th-magnitude stars designated NGC 7772. All of the stars in this cluster occupy an area of the "window" smaller than a star dot on our map. A telescope with a larger diameter should show even fainter stars. Altogether there are hundreds of thousands of stars in the wedge, reaching out to the "bottom" of the Galaxy. "Bottom," by the way, refers to earth's south—a little bit of northern chauvinism.

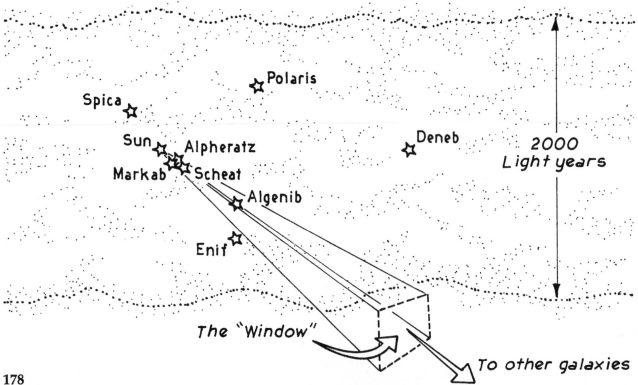

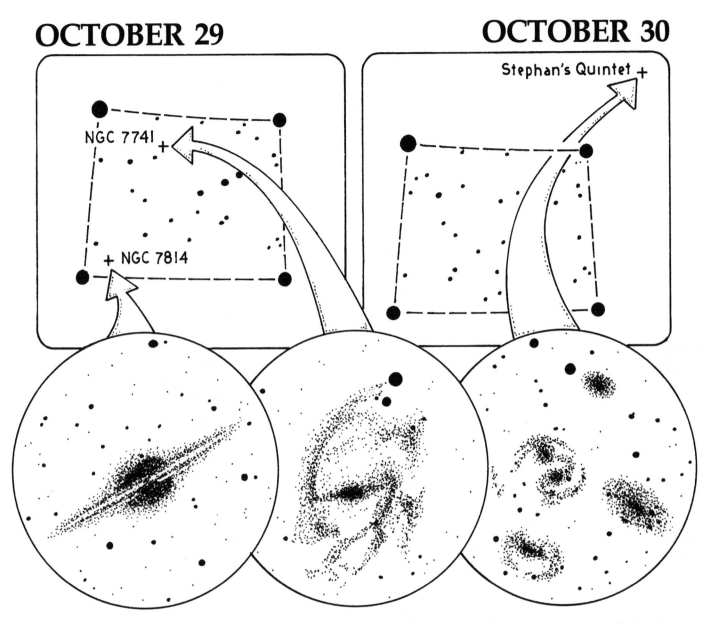

29th: About 2000 light years through our Pegasus "window," we come to the edge of the star distribution that is the Milky Way Galaxy. But that is not the end of things to see through the "window." Beyond the Milky Way there are other galaxies. Two striking galaxies that can be seen through the Pegasus "window" are shown above. NGC 7814 is a spiral galaxy seen edge-on. The great central bulge of the nucleus is prominent. The lanes of dust and gas that are characteristic of the arms of spiral galaxies are also prominent in our view of this remarkable object. NGC 7741 is a typical "barred" spiral galaxy. From the central nucleus two major arms extend straight outward like spokes of a wheel for half the radius of the galaxy. More or less typical spiral arms unwind from the end of these "bars." It is not known why the barred spirals have their curious shape. These two galaxies lie tens of millions of light years from the earth. Beyond them lie even more distant galaxies.

30th: The little group of galaxies I have shown above is actually a bit outside our "window frame," but beautifully represents our deeper vista into space. Stephan's Quintet of five galaxies could be as far as 400 million light years away. The true distance has been hotly debated. There now seems to be a consensus that the largest member of the group is not physically related to the others, but lies in the foreground and is only coincidentally along the same line of sight.

OCTOBER 31

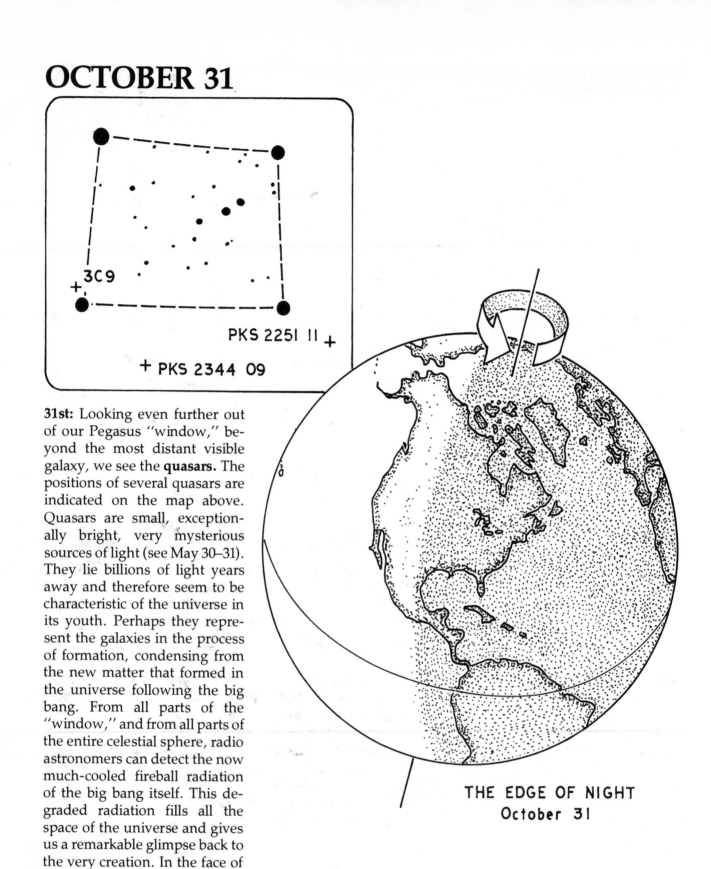

3C9

PKS 2251 11

+ PKS 2344 09

31st: Looking even further out of our Pegasus "window," beyond the most distant visible galaxy, we see the **quasars.** The positions of several quasars are indicated on the map above. Quasars are small, exceptionally bright, very mysterious sources of light (see May 30–31). They lie billions of light years away and therefore seem to be characteristic of the universe in its youth. Perhaps they represent the galaxies in the process of formation, condensing from the new matter that formed in the universe following the big bang. From all parts of the "window," and from all parts of the entire celestial sphere, radio astronomers can detect the now much-cooled fireball radiation of the big bang itself. This degraded radiation fills all the space of the universe and gives us a remarkable glimpse back to the very creation. In the face of such grand and mysterious events, let us remember the words of the physicist Michael Faraday: "Nothing is too wonderful to be true."

THE EDGE OF NIGHT
October 31

180

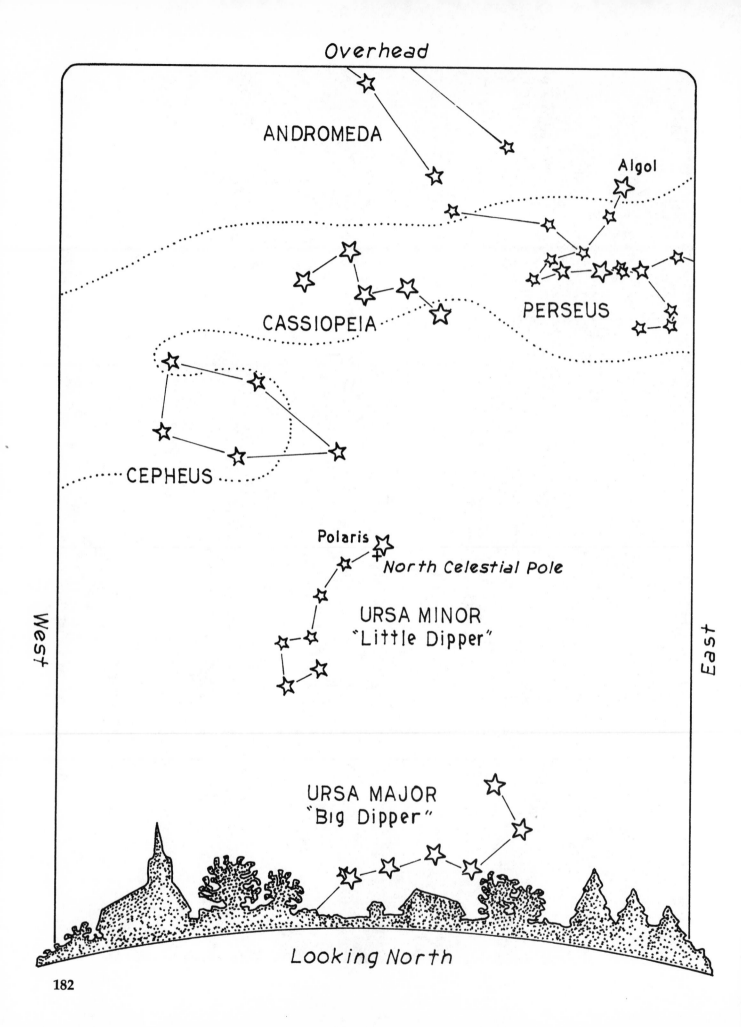

Overhead

ANDROMEDA

Algol

PERSEUS

CASSIOPEIA

CEPHEUS

Polaris
North Celestial Pole

URSA MINOR
"Little Dipper"

West

East

URSA MAJOR
"Big Dipper"

Looking North

Overhead

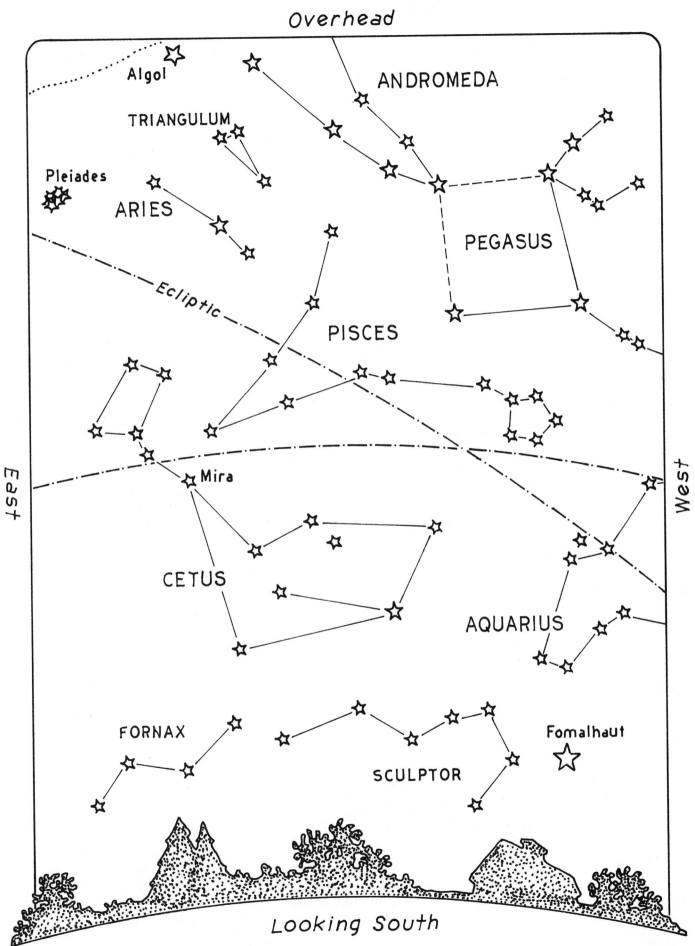

ANDROMEDA

Algol

TRIANGULUM

Pleiades

ARIES

Ecliptic

PISCES

PEGASUS

East

Mira

CETUS

AQUARIUS

West

FORNAX

SCULPTOR

Fomalhaut

Looking South

183

NOVEMBER 1

NOVEMBER 2

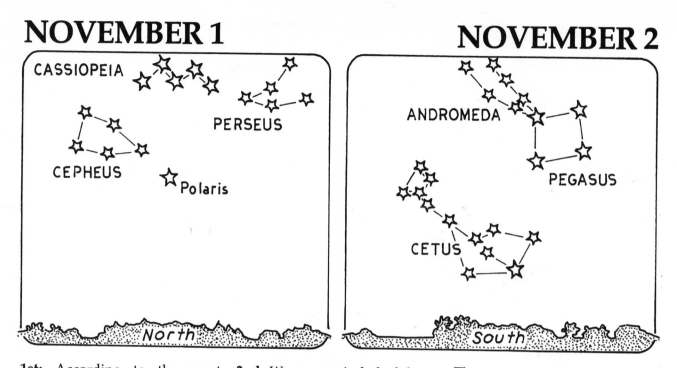

1st: According to the most widely held contemporary theory, the universe began about 15 billion years ago in a single colossal explosion that created not only the "stuff" of the universe, but space and time as well. What came before the "big bang," if that question has any meaning, is locked up in the mystery of infinity. At the first instant, all of the matter of the universe was compressed to an infinite density, a sort of a super **black hole.** During the first few seconds of the expansion, the temperature was greater than a trillion degrees. After about a minute-and-a-half the temperature of the fireball had cooled to a billion degrees. It took a million more years for the universe to cool to something akin to the temperature of the sun's surface. Today the background temperature is about 3° above absolute zero (or −270°C). It is the original fireball radiation, cooled to −270°C, that radio astronomers now detect in every part of the sky.

2nd: We are reminded of those far-off times and mind-boggling events by the cooling of the season. Nights now begin to take on winter's chill. On the eastern edge of the star map on the previous page are the Pleiades, with which we began the year. In their rising we anticipate the completion of the cycle of the sky and the subsequent warming of the earth. Whether the expansion and cooling of the universe itself will one day reverse has yet to be determined. Our present-day story of creation is based on esoteric theories of physics and subtle observations with complex instruments. Our ancestors had a less abstract story of a creation and took a more familiar mythological approach. The sky overhead this month includes the full cast of characters for one of their most fascinating inventions. Cepheus, King of Ethiopia, was married to Cassiopeia. She bore him a beautiful daughter Andromeda. Cassiopeia boasted that her daughter was more beautiful than the sea nymphs.

The sea nymphs, hearing of the boast, were jealous. They protested to Poseidon, god of the sea. Poseidon agreed to punish Cassiopeia for her boastfulness by sending the sea monster Cetus to ravage the coast of Ethiopia. The monster began its terrible work. Cepheus consulted an oracle as to how to stop the devastation. He was advised that the only way the gods could be appeased was the sacrifice of his daughter. Andromeda was to be chained to the coastline for Cetus to devour. Cepheus was torn between love for his daughter and concern for his people. In the end he made the choice and had Andromeda chained to a sea rock. But even as Cetus approached the doomed girl, the hero Perseus appeared on his winged horse Pegasus. Perseus looked at Andromeda and was smitten by her beauty. He slew Cetus and claimed the hand of Andromeda. They lived a long and happy life together and, ultimately, all of the characters in the story were given places in the sky.

NOVEMBER 3

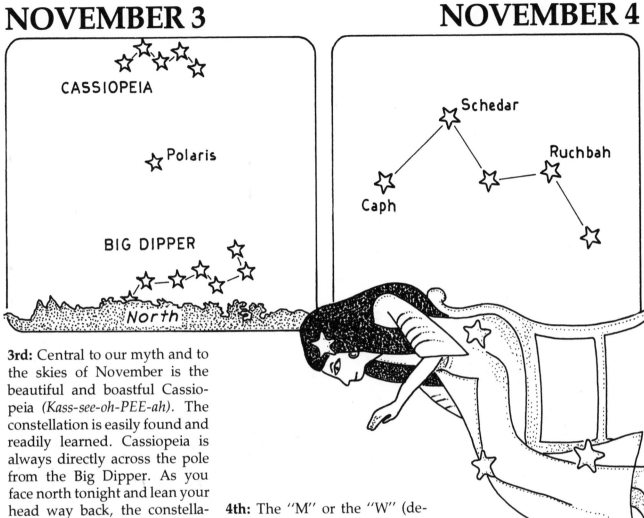

CASSIOPEIA

☆ Polaris

BIG DIPPER

North

3rd: Central to our myth and to the skies of November is the beautiful and boastful Cassiopeia *(Kass-see-oh-PEE-ah)*. The constellation is easily found and readily learned. Cassiopeia is always directly across the pole from the Big Dipper. As you face north tonight and lean your head way back, the constellation will look like an "M" of five bright stars. When Cassiopeia changes places with the Dipper in 6 months' time, it will stand as a "W" on the northern horizon. It takes a considerable exercise of the imagination to see a woman seated on a chair. Sometimes it is only the throne itself that is associated with the five stars. If indeed we do choose to see the boastful queen upon her chair, it is difficult to imagine how she keeps from falling out at some time of the year. Perhaps by the same celestial magic that keeps the contents of the Big Dipper from pouring out when the positions of the constellations are reversed.

NOVEMBER 4

Schedar

Ruchbah

Caph

4th: The "M" or the "W" (depending on the season) is certainly one of the most familiar star groups in the sky. For observers at mid-northern latitudes the constellation is **circumpolar,** which means it is close enough to the pole so that it never sets. The queen would be pleased with her present position near the **zenith,** high on her royal throne, her family clustered about her. The brightest star in Cassiopeia is Schedar *(SHED-der)*. The name means "breast." Almost equally bright is Caph *(kaff)*, a star only 45 light years away and very similar to Altair and Procyon in size and temperature. Ruchbah *(RUCK-bah)* means "knee." As you can see, the way I have drawn the constellation does not correspond with the way the queen was pictured by those who named the stars. I have chosen to draw Cassiopeia in the posture I learned as a boy, long before I knew the star names or their meanings. The habits of a lifetime are hard to break. The very uncommon names of the remaining two stars of the familiar five have dubious meanings and do not help us seat the queen upon her throne. These last two stars are usually designated by Greek letters only, *gamma* and *epsilon.*

NOVEMBER 5

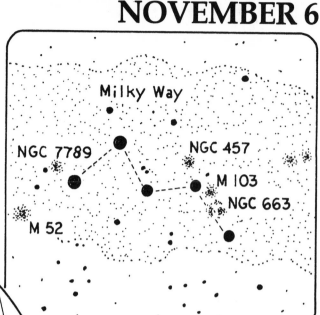

NOVEMBER 6

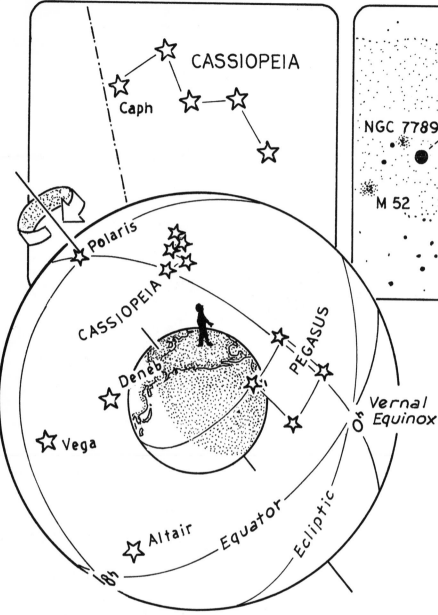

5th: Caph in Cassiopeia (Beta Cassiopeiae) is a useful guide to the web of celestial coordinates which the imagination must impose on the sky. The line which passes from Polaris through Caph and on down through the two eastern stars of the Great Square of Pegasus is the zero-hour line of right ascension. This is the "prime meridian" of the sky, and intersects the celestial equator at the **vernal equinox.** Hours of right ascension are measured eastward around the equator from that point. The right ascension of Caph is therefore close to zero (R.A. 0h 8m, dec. +59°). If you have a clear sky this evening, put yourself in the position of the observer in the drawing above. With your arm, see if you can trace out the "prime meridian," the equator, and the ecliptic on the black slate of the sky. You have all the hints you need.

6th: It is near Cassiopeia that the stream of the Milky Way sweeps closest to the north pole of the sky. This is a part of the stream where the "milk" runs thick. The constellation is wonderfully rich in star clusters, those **open galactic clusters** of young stars that are typical of the "gassy" arms of the Galaxy. Here in Cassiopeia new stars are born, condensing from the nebulous matter of the spiral arms. And here massive blue stars quickly exhaust their energy and die violent and explosive deaths. The map above shows some of the jewellike clusters that are accessible to the observer through a telescope. The typical population of one of these clusters might be a thousand stars, all twinkling in a volume of space no larger than 30 light years across. The age of the clusters can be determined from an **H-R diagram** (see Mar. 19–20). The beautiful cluster M 52, for example, appears to be only about 20 million years old. The clusters are thousands of light years away.

186

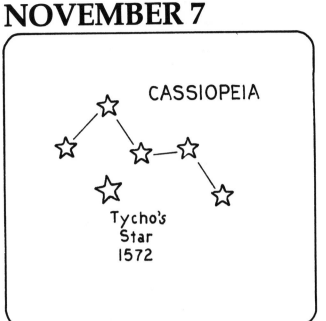

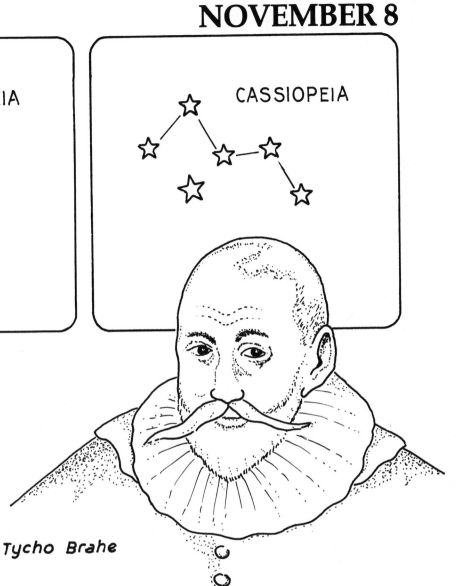

Tycho Brahe

7th: On November 11 in the year 1572, the great Renaissance astronomer Tycho Brahe, as was his custom, contemplated the stars in the clear evening sky. Suddenly he noticed, almost directly over his head, a new star, surpassing in brilliance all other stars in the sky. Tycho was a consummate mapper of the heavens, and knew immediately that no star, not even the most faint, had previously occupied that part of the sky. Such changes in the heavens were unheard of, and Tycho briefly doubted the evidence of his own eyes. He called to his friends and assistants and they too verified the "new star," or nova. It was a miracle, Tycho believed, the first of its kind since the Creation. He was wrong, of course. Earlier **supernovae** had been recorded, such as the explosion that produced the Crab Nebula in 1054, but they were unknown to Tycho. No matter, the brilliant new star in Cassiopeia was a wonder to be admired in any age, and has not been rivaled since.

8th: There are written records of two certain earlier **supernovae,** including the one which produced the Crab, and a larger number of **novae.** Chinese records are a rich source of information concerning these "guest stars," as the Chinese called them. Other novae and supernovae had undoubtedly been observed but not recorded by our prehistoric ancestors, such as the furious blast that produced the Veil Nebula in Cygnus tens of thousands of years ago. Presumably a supernova of the kind observed by Tycho oc-

curs every few hundred years in a typical galaxy. What Tycho saw in Cassiopeia was the violent and almost totally self-destructive death of a massive star. That event occurred 400 years ago. Another occurred a few years later in 1604, the supernova we call Kepler's Star. We are probably overdue for another. Perhaps it will be on this starry night, and, like Tycho, you and I can be witnesses to the "miracle" of a new star. Ours is a less credulous age, but let us hope we have not lost our capacity to be astonished.

NOVEMBER 9

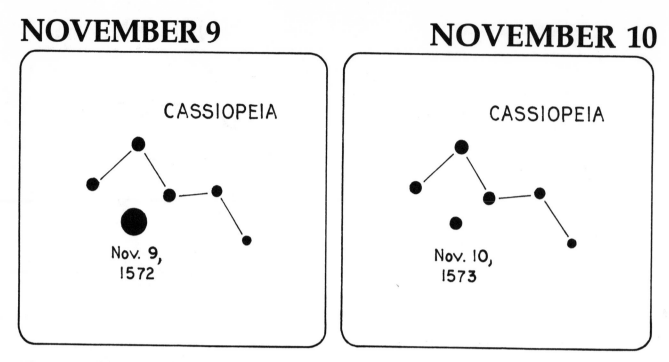

CASSIOPEIA

Nov. 9,
1572

NOVEMBER 10

CASSIOPEIA

Nov. 10,
1573

9th: At its brightest, Tycho's Star was more brilliant than Venus and could even be seen in daylight. Slowly it faded, changing color from white to yellow to reddish, becoming invisible to the naked eye in about 18 months. Modern studies indicate that the star that destroyed itself was about 10,000 light years from the earth. It may have had a peak luminosity of hundreds of millions of times that of the sun.

10th: The debris from the death of Tycho's Star is still expanding and will ultimately become part of the diffuse clouds of dust and gas that fill the spiral arms of the Galaxy. It is now believed by many astrophysicists that the heavier elements now present in the universe (such as carbon, oxygen, iron, and others) are created from the lighter elements (hydrogen and helium) in the interiors of stars. The process, known as thermonu-

clear fusion, takes place during the period of long, slow burning on the **main sequence** and in the even more energetic conditions of a **supernova** explosion. The newly created heavy elements are dispersed by the explosion. If that is so, then the very atoms of which you and I are made were "cooked up" in stars that lived and died long before the solar system had its beginning. We may be literally made of "stardust."

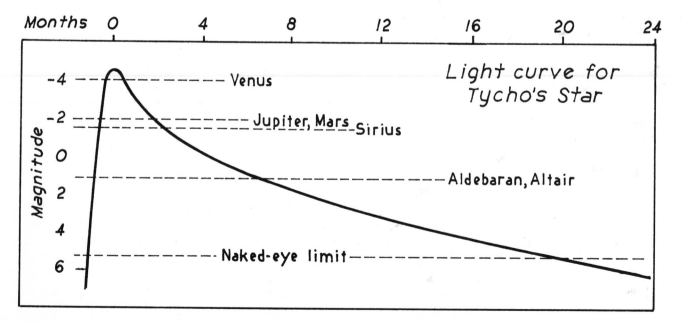

Months

Light curve for Tycho's Star

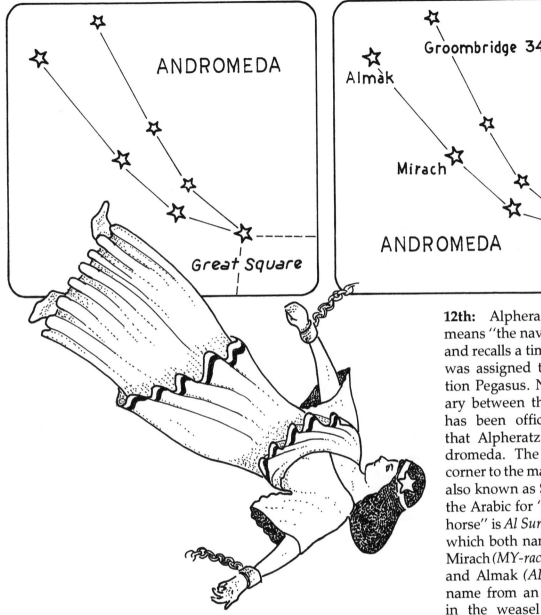

ANDROMEDA

Great Square

Almak

Groombridge 34

Mirach

Alpheratz

ANDROMEDA

12th: Alpheratz (al-FEE-rats) means "the navel of the horse," and recalls a time when the star was assigned to the constellation Pegasus. Now the boundary between the constellations has been officially drawn so that Alpheratz is part of Andromeda. The square loses a corner to the maiden. The star is also known as Sirrah. Actually, the Arabic for "the navel of the horse" is *Al Surrat al Faras*, from which both names are derived. Mirach *(MY-rack)* means "loins" and Almak *(AL-mack)* takes its name from an Arabian animal in the weasel family. Rather than as a weasel, you will better think of the star as Andromeda's foot. Also in Andromeda, but invisible to the unaided eye, is the tiny **binary** system Groombridge 34. This little pair of stars is our 16th nearest neighbor, two of the host of nearby red dwarfs. Three-quarters of our 50 nearest neighbors are small red dwarfs. Groombridge was an astronomer who compiled a catalogue of **circumpolar stars** in 1838.

11th: Andromeda *(an-DROM-eh-dah)*, the chained woman, was the beautiful daughter of Cassiopeia who was offered in sacrifice to the sea monster. On the old star maps she is often shown chained to a rock on the seacoast, awaiting her terrible fate in the jaws of Cetus. In the sky she is separated from the ravaging monster by the constellation Pisces the Fish. This evening you will find Andromeda directly over your head. This is not a particularly conspicuous group of stars. Your best bet is to find the Great Square of Pegasus and then look for a long lazy "vee" of stars extending from the northeast corner of the square. In every case, the stars along the southern arm of the "vee" are brighter than those on the northern branch. If you think of Andromeda in a flowing dress, it is not hard to associate the "vee" of stars with her arching body.

NOVEMBER 13

NOVEMBER 14

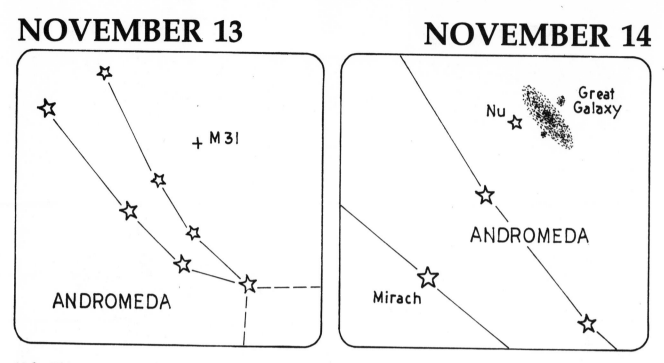

13th: The jewel of Andromeda is the *Great Nebula*, M 31. It is one of the few objects on Messier's list that can be seen with the unaided eye. On a clear moonless night you may see the bright central part of M 31 as a hazy patch of light. As we shall see, it is the most distant object you will ever see without a telescope.

14th: Long-exposure photographs reveal M 31 to be a giant spiral galaxy, filling a part of the sky almost four times broader than the full moon! The outer parts of the spiral are very faint. We see the galaxy very close to edge-on, so its spiral structure is not obvious. Nearby are two small satellite galaxies, M 32 and NGC 205. The group is at a distance of 2 million light years, which puts it far beyond the stars of our Milky Way. The light that enters your eye left the Andromeda spiral at about the time the species *Homo sapiens* first appeared on this planet. The great distance and true nature of the nebula were first recognized in 1923, when Hubble found **cepheid variables** among the nebula's myriad stars.

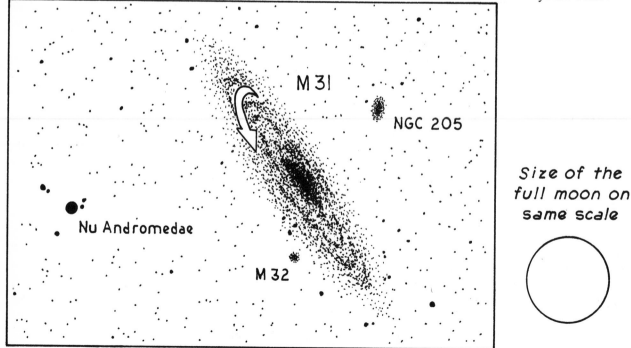

Size of the full moon on same scale

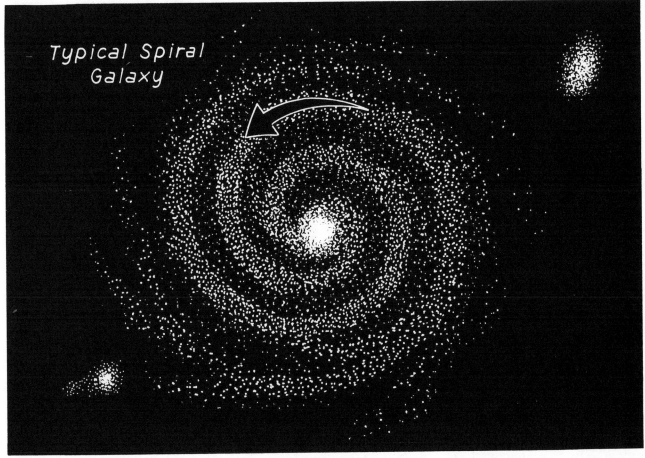

Typical Spiral Galaxy

15th: Only in recent times did astrophotography reveal that the hazy glow of the *Great Andromeda Nebula* was actually the light of hundreds of billions of individual stars in a splendid spiral structure. At a distance of 2 million light years, only the very brightest stars show up on photographs made with the world's largest telescopes. A star the size of our sun, if in the Andromeda Galaxy, would be totally invisible. Fortunately, **cepheid variable** stars are luminous giants. It was the cepheids that Edwin Hubble used to determine the distance to the Andromeda Galaxy (see Sept. 16–20). Picking out these variable stars from thousands of pinpricks of light on a series of photographic plates was no easy task.

16th: The *Great Galaxy in Andromeda* is one of the largest galaxies known, more than half again as large as the Milky Way Galaxy. The nucleus is an ellipsoidal mass of old red and yellow giant stars. The spiral arms are rich with dust and gas and contain many hot young blue stars. The entire aggregate of several hundred billion suns rotates about the central hub in from 10 to 200 million years, depending on the distance to the center. Like the Milky Way Galaxy, M 31 has a halo of globular clusters and several satellite galaxies. The drawing above could represent the Great Andromeda Galaxy or our own Milky Way. A photograph of the Milky Way from the Andromeda Galaxy would presumably look very similar (except for scale) to an earth photograph of M 31. Even the orientation of the disks of the galaxies to the line of sight is the same. It is worth wondering if such a photograph exists on the planet of a sunlike star in M 31. It is hard to believe that in such a myriad of worlds there is not life. I have often made a model of a spiral galaxy on the floor with a box of salt. It would take *10,000 boxes of salt* to have as many grains in the model as there are stars in the Andromeda Galaxy! For the scale to be approximately right (remember galaxies are mostly empty space) the 10,000 boxes of salt would have to be spread out over an area the size of the moon's orbit.

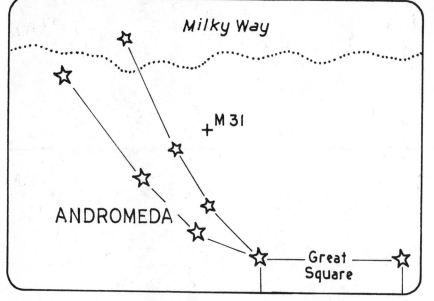

18th: Galaxies tend to come in clumps. The Milky Way Galaxy belongs to a cluster called the **Local Group.** There are more than 20 galaxies in the Local Group. Of these, the Milky Way and the Andromeda spiral are the two giants. There is only one other spiral, the beautiful face-on Pinwheel Galaxy in the constellation Triangulum (see Nov. 21). The Pinwheel is little more than half the size of the Milky Way, a more typical size for spiral galaxies. There are also a dozen or so elliptical galaxies, small egg-shaped swarms of stars such as the two companions of M 31. There are several irregular galaxies, ragged aggregates of stars such as the Milky Way's close companions, the Magellanic Clouds. The obscuring matter of the Milky Way may prevent us from seeing still other nearby companions in this corner of the universe. The member galaxies of the Local Group move at random within the group. The Andromeda Galaxy, for example, is approaching us at a velocity of 22 miles per second.

17th: If the *Andromeda Galaxy* were the size of this book, the earth would shrink to the size of an electron, an insignificant speck in a sea of stars. But that tiny electron is our blue planet, and our platform for viewing the universe. In the drawing below I have exaggerated the size of the earth to show its position this evening relative to the Galaxy. The plane of the earth's orbit around the sun is inclined by about 60° to the disk of the Milky Way. The sun is a larger star than the majority of the stars in the Galaxy, but still tiny compared to the red and blue giants that show up on photographs of M 31. Don't forget that the stars that make up the *constellation* of Andromeda are nearby stars in our Milky Way Galaxy. We look out through these stars, across 2 million light years of empty space to the Great Galaxy of Andromeda.

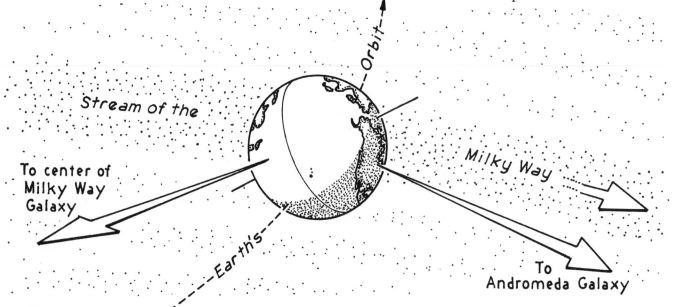

NOVEMBER 18

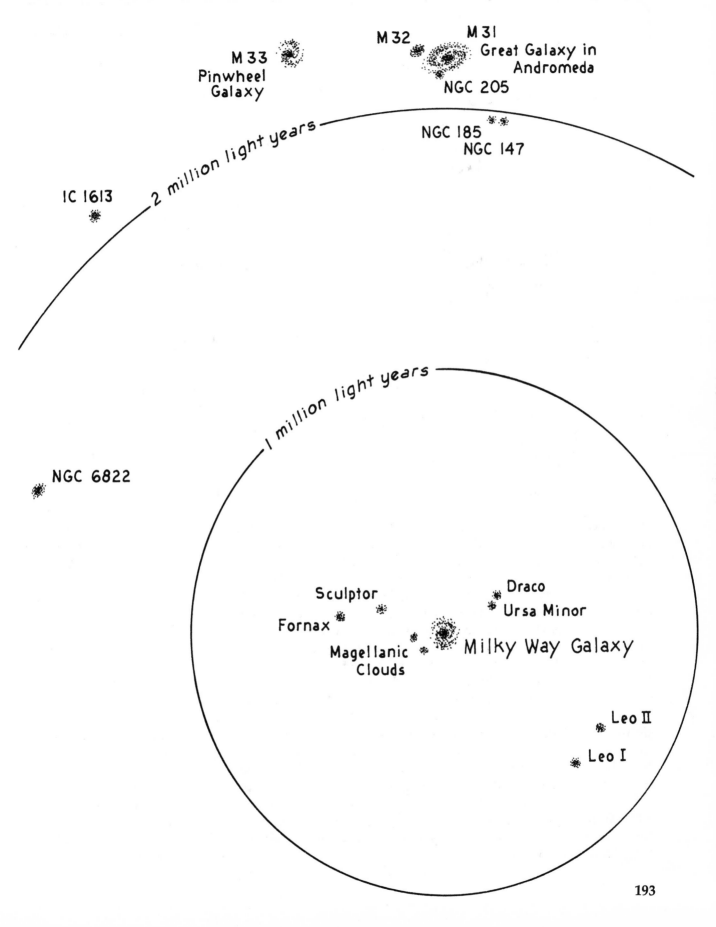

M 33
Pinwheel
Galaxy

M 32

M 31
Great Galaxy in
Andromeda

NGC 205

2 million light years

NGC 185
NGC 147

IC 1613

1 million light years

NGC 6822

Sculptor

Fornax

Draco
Ursa Minor

Magellanic
Clouds

Milky Way Galaxy

Leo II

Leo I

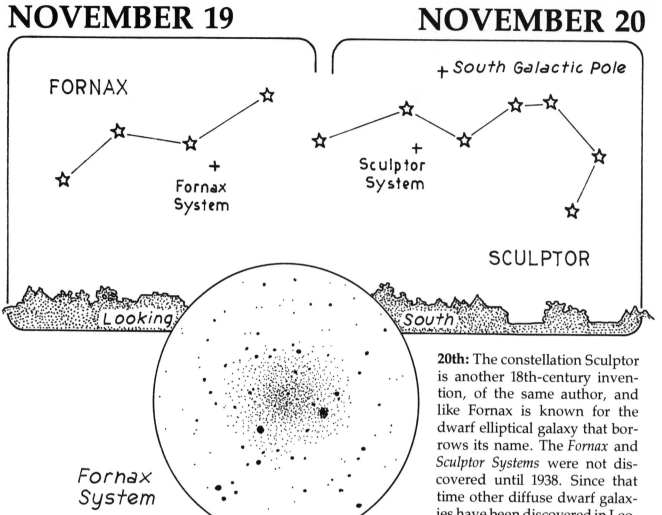

FORNAX

+ Fornax System

+ South Galactic Pole

Sculptor System +

SCULPTOR

Looking   South

Fornax System

19th: The constellations Fornax and Sculptor have given their names to two of our closest galactic neighbors. This is perhaps their most conspicuous claim to our attention. Otherwise the two constellations, low in our southern sky, have little to recommend them. Neither has a star brighter than the 4th magnitude. Fornax means "furnace," a rather unlovely 18th-century name for this faint group of stars. Within the bounds of Fornax is the ghostly globe of stars known as the *Fornax System.* The object is about 600,000 light years away, or three times further than the Magellanic Clouds. It resembles

a globular cluster that has been "puffed out" to wraithlike thinness. The Fornax galaxy covers an area of the sky larger than the moon, but is so faint and diffuse that it is only accessible to major observatory telescopes. These wispy elliptical galaxies may in fact be very common in the universe, but they would be impossible to detect unless very near. A citizen of the Fornax System would have no such difficulty detecting the presence of our galaxy. Indeed, all other things being equal, the Milky Way Galaxy would be a splendid naked-eye object in the night sky of a Fornax System planet.

20th: The constellation Sculptor is another 18th-century invention, of the same author, and like Fornax is known for the dwarf elliptical galaxy that borrows its name. The *Fornax* and *Sculptor Systems* were not discovered until 1938. Since that time other diffuse dwarf galaxies have been discovered in Leo, Draco, and Ursa Minor, all members of the **Local Group.** Also within the borders of the constellation Sculptor is the south **galactic pole.** When we look toward Sculptor we are looking straight down out of the disk of the Galaxy. This accounts for the paucity of bright stars in this part of the sky. It also makes it easier to see faint galaxies such as the Fornax and Sculptor Systems. Galaxies, by the way, are classified as elliptical, lenticular (lens-shaped), spiral, barred spiral, and irregular. The symbols used to designate these types are E, SO, S, SB, and Ir. Sometimes a "p" is added to indicate "peculiar." The different types probably do not represent stages in the evolution of galaxies.

NOVEMBER 21

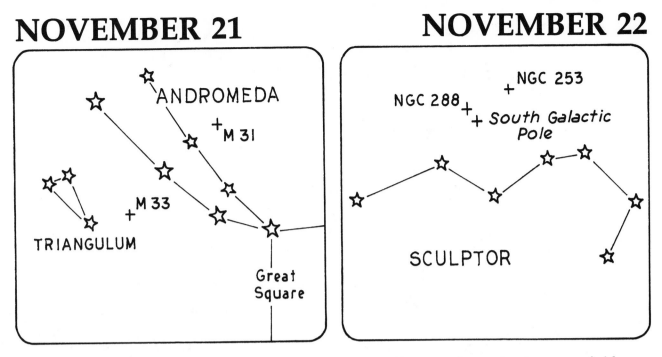

NOVEMBER 22

21st: Close by the Great Andromeda Galaxy is the only other spiral in the **Local Group,** M 33. The *Pinwheel Galaxy* is in the constellation Triangulum ("Triangle"). The many-armed face-on spiral lives up to its name. You might see it with binoculars, but don't expect to see the pinwheel arms. Galaxies such as this give an impression of spinning, but the periods of rotation are too long for us to observe movement.

22nd: Beyond the **Local Group** of galaxies, the next significant clusters are in Ursa Major and Sculptor. The *Sculptor Group* is 6 or 7 million light years distant, compared to 2 million light years for the Andromeda Galaxy and the Pinwheel. The most prominent member of the Sculptor Group is the spiral NGC 253, an easy object for small telescopes, especially for southern observers. In the drawing below the spiral is

shown in the same field as a **globular cluster** (NGC 288) of the Milky Way Galaxy. It is a striking conjunction of a relatively nearby and very distant object. The stars shown are even nearer than the globular cluster. They lie between us and the flat edge of the Milky Way spiral, all closer than a thousand light years. The nearby groups of galaxies are part of a "super-cluster" with a center toward the constellation Virgo.

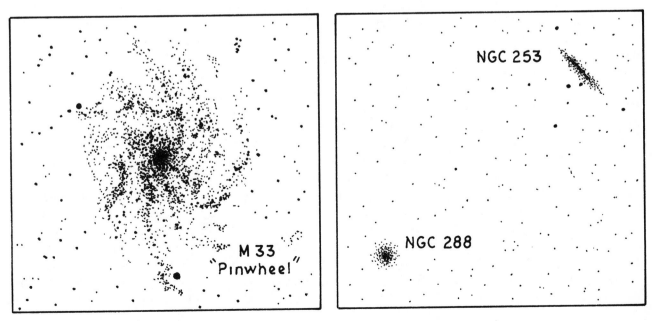

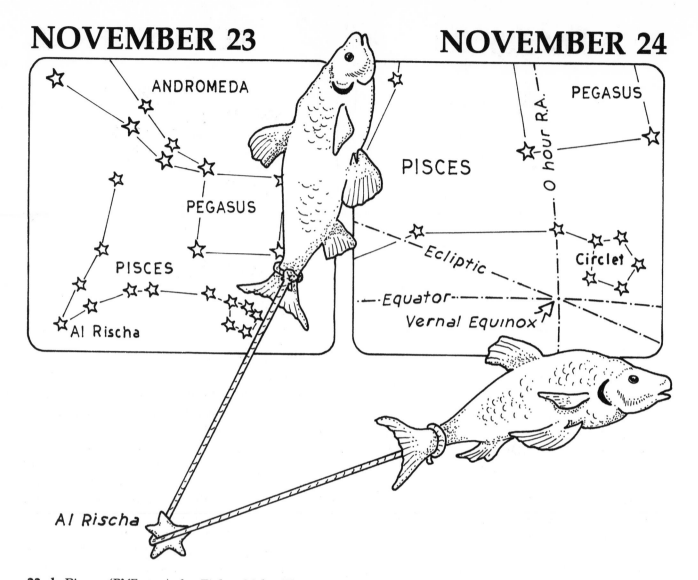

23rd: Pisces (*PYE-sees*) the Fish is another of the watery constellations common in this part of the sky. The association of these faint stars with fish is very ancient. Babylonians, Persians, and Arabs all saw the same image in the sky. There are no bright stars in the constellation, and you will want a perfect night to trace the open "vee" of stars that represents two fish and the cord that ties them together. The most prominent star in the group is Al Rischa (*al-ri-SHAY*) at the vertex of the "vee." The name means "the cord." Our familiarity with the constellation rests entirely on its inclusion in the **zodiac.**

24th: The sun is in Pisces in March and April. Thousands of years ago, when astrology was invented, the sun was in Aries during late March and early April, and in particular on the day of the **spring equinox.** Aries, accordingly, was at the beginning of the list of constellations of the **zodiac.** The equinox was called "the first point of Aries," and most astrologers still make their calculations on the ancient basis. If you were born "under the sign of Pisces," that means that the sun was in Pisces on your birthday—two thousand years ago! Nowadays the sun would stand in the constellation Aquarius on the day

you were born, a fact of remarkably little concern to astrologers. Since the birth of astrology, the slow **wobble** of the earth's axis has caused the spring equinox to move into (and almost across) the constellation Pisces. It now lies near the head of the western fish, a little pentagon of stars known as the *circlet.* This is the point which is our "Greenwich" for measuring **right ascension.** The spring equinox moved into Pisces at the beginning of the Christian era. The Christian era has been called, therefore, the "Age of the Fish," and the fish has remained an important symbol in that religion.

196

NOVEMBER 25

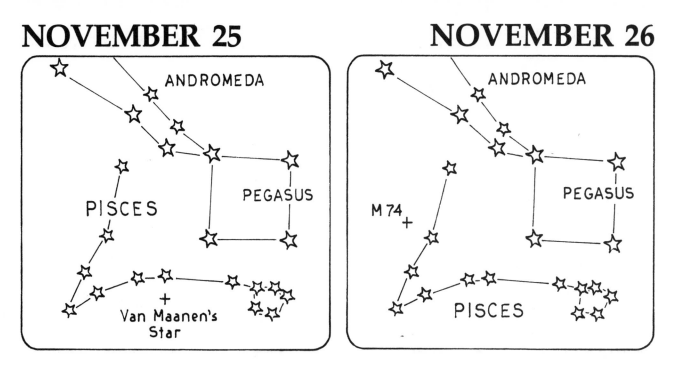

NOVEMBER 26

25th: Near the tail of the western fish is Van Maanen's Star, one of our nearest neighbors in space, and except for the companions of Sirius and Procyon, the nearest **white dwarf.** It is about 14 light years away. Van Maanen's star is the collapsed and dying remnant of a star that was once larger than the sun. Gravity has squeezed a mass about that of the sun's into a volume that is the size of the earth. So dense is the matter of this star that a single teaspoonful would weigh as much as an automobile!

26th: Pisces offers an excellent opportunity to continue our survey of galaxies deeper into space. M 74 is another beautiful face-on spiral galaxy, of about the same size and structure as the Milky Way. It is somewhat more tightly wound than the Pinwheel. The galaxy lies about 30 million light years from earth, or 15 times further than the Great Galaxy in Andromeda. From its spectrum (see Nov. 30) we can deduce that it is moving away from us with a velocity of 400 miles per second. This movement of re-

cession "stretches" the wavelengths of the light that reaches us. This so-called reddening of the galaxy's light is the clue that gives us the velocity. Beyond the **Local Group,** *all of the galaxies are in recession.* This strange fact does not mean that we are special, but that the galaxies (including our own) are moving away from one another. It was the discovery of this remarkable fact that led to the concept of an *expanding universe.* The recession of the galaxies was announced by Edwin Hubble in 1929.

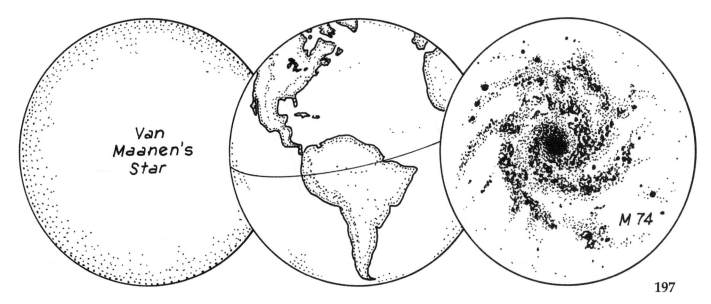

NOVEMBER 27

NOVEMBER 28

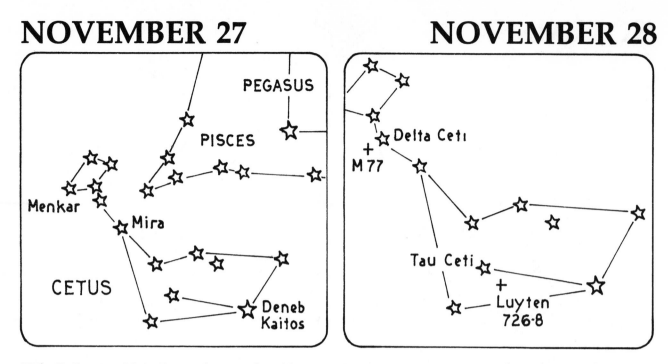

27th: Following H.A. Rey, who has written delightful books on the stars, I like to see Cetus the Whale in the posture suggested by the map. This completely ignores the names of Menkar ("nostril") and Deneb Kaitus ("tail of Cetus"). But it is not Menkar and Deneb Kaitus for which the constellation is best known. Mira ("wonderful") is the special attraction of the whale. Mira was the first variable star to be clearly recognized as such. It comes and goes, appearing and disappearing with a wonderful regularity.

28th: Also in Cetus are two more of our nearest neighbors. Luyten 726-8 is the 6th closest star system to the earth, only slightly more distant than Sirius. It is a **red dwarf** binary system, invisible to the naked eye. The two tiny stars have masses less than almost any other stars. These miniscule objects, not much more massive than Jupiter, are probably common in the Galaxy. But we would not see them if they were not "next door." Tau Ceti is the 18th nearest star system, 12 light years away. It is a naked-eye star, and is very similar to our own sun. Because of this similarity, Tau Ceti is considered a likely place to search for intelligent extraterrestrial life. In fact, the star has been carefully surveyed by radio astronomers for signals of an intelligent origin. Part of the problem, of course, is defining "intelligence." What kind of message should we look for? Certainly, evolution on distant planets might have led to utterly alien ways of "thinking." In any case, the search for signals has been unsuccessful.

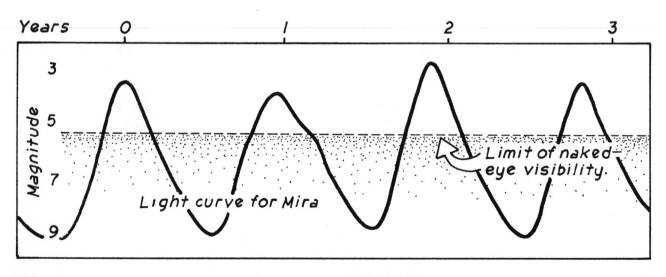

NOVEMBER 29

NOVEMBER 30

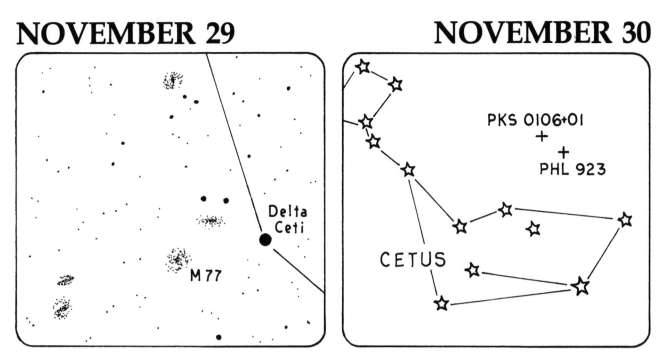

29th: Close by Delta Ceti is a cluster of distant galaxies. M 77, a nebula on Messier's list, is the most prominent member of the group. It is a face-on spiral that resembles a three-bladed propeller. M 77 was one of the first galaxies for which a **red shift** was observed. When a source of sound or light moves toward or away from an observer, the pitch and wavelength of the sound or light are modified. If the source moves away from the observer the wavelengths are "stretched out" and the pitch lowered. If the source moves toward the observer the wavelengths are "compressed" and the pitch raised. You have heard this effect, called the *Doppler effect,* in the sound a truck makes as it passes on the highway—high pitch in approach, low pitch in recession. Similarly, the wavelengths of light in a receding star or galaxy are lengthened. Since the long wavelengths of visible light are red, we say the light of the object is *red-shifted*. Light from an approaching source is *blueshifted*. The degree of shift is related to the velocity.

30th: In an expanding universe, more distant objects will have greater velocities of recession. Once the rate of expansion is known, **red shifts** can be used to determine distance. The most distant objects in the universe are the **quasars.** There are notable quasars in Cetus. PHL 923 lies 10 billion light years away. The red shift of PKS 0106+01 indicates a velocity of recession of 8/10 the speed of light. That places the quasar nearly 15 billion light years from earth, and back in time to the earliest era of the universe!

Red shift of hydrogen lines in quasar spectrum

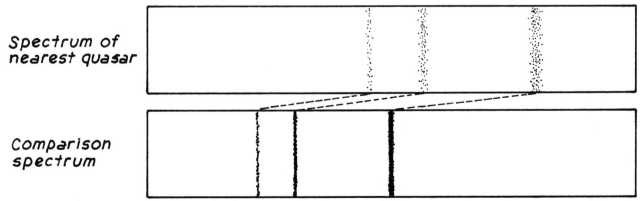

THE EDGE OF NIGHT
November 30

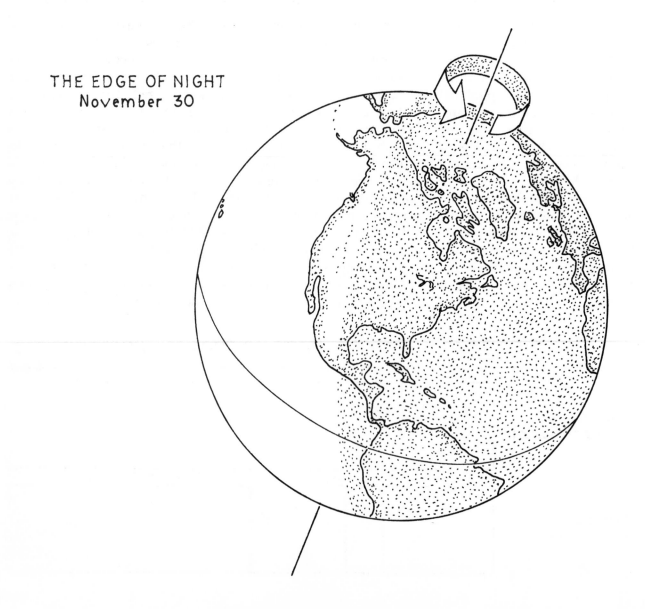

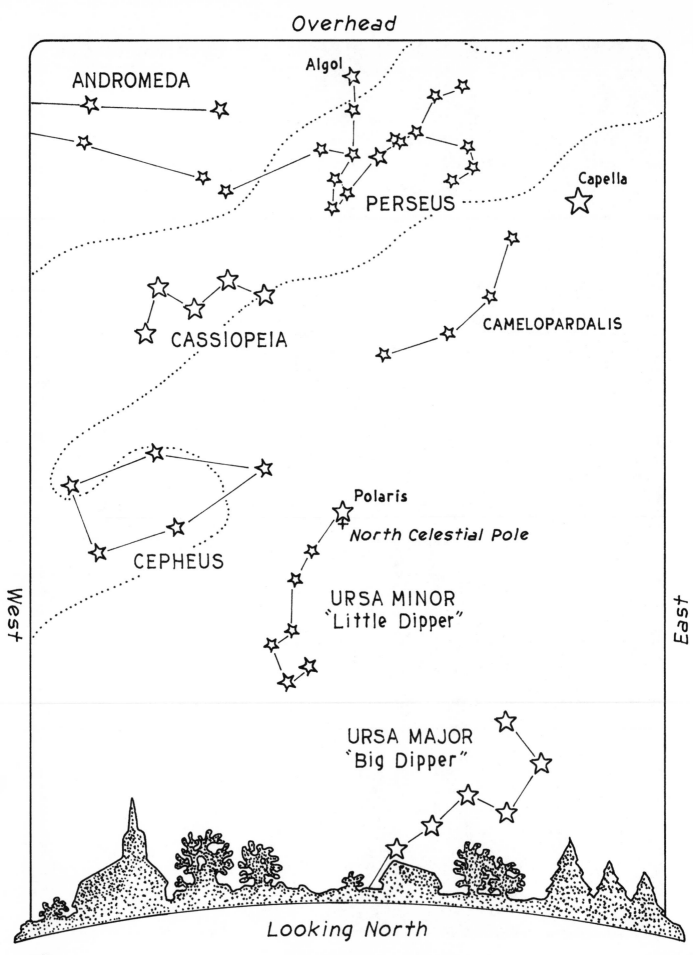

Overhead

ANDROMEDA

Algol

PERSEUS

Capella

CASSIOPEIA

CAMELOPARDALIS

CEPHEUS

Polaris

North Celestial Pole

URSA MINOR
"Little Dipper"

URSA MAJOR
"Big Dipper"

West

East

Looking North

202

DECEMBER 1

DECEMBER 2

1st: The cycle of the seasons is almost complete. As the sun sets this evening you will see Orion rising in the east. The Bull is already high in the sky, and his red eye gleams above the east point on your compass as the sky darkens. In the northeast Capella broods watchfully over her kids. Seeing these stars again reminds us of the brilliance of winter constellations, and whets our appetite for returning once more to the stars of January. Meanwhile, if you turn toward the west, you can watch the curtain ringing down on the stars of summer. Vega and Altair slide together toward the horizon, followed closely by Deneb, still affirming the theorem of Pythagoras. Overhead in the south we remain at intermission between the two lively acts of the celestial drama. So this month let's turn once more to the north. We shall look chiefly at the constellation Perseus. The rescue on the rocky coast of Ethiopia has been accomplished. Cetus sinks to his watery grave. And beautiful Andromeda has come firmly and forever between the slayer of Medusa and his steed.

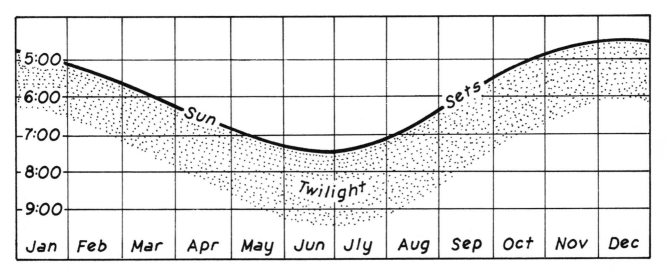

2nd: December is the month of the earliest sunsets and the longest evenings of star watching. Like a bead on a wire, the sun slides down the **ecliptic** toward its lowest point. It sets now well south of west and will creep a little further in that direction before the **winter solstice.** The exact clock time the sun disappears from view this evening will depend on your latitude, the "flatness" of your horizon, and where you live within your time zone. But by six-thirty the sky should be dark and the winter stars shining brightly. At mid-northern latitudes, the actual date of the earliest sunset is near the beginning of the second week of December (the shortest interval between sunrise and sunset occurs at the beginning of the fourth week). Of course, the sky does not darken at the instant of sunset. The atmosphere scatters the sun's light and the sky remains illuminated for an hour or more until the sun has dropped a handspan below the horizon. The length of twilight depends upon latitude and season.

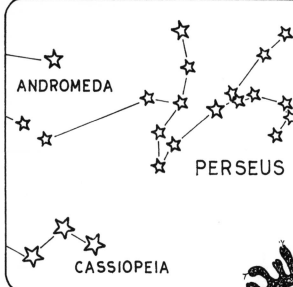

ANDROMEDA

PERSEUS

CASSIOPEIA

Algol

PERSEUS

Mirfak

3rd: Our system of constellations had its origins in the river valleys of the Near East and was later embellished by the Greeks. From the latitudes of Mesopotamia and the eastern Mediterranean only those parts of the celestial sphere can be observed with declinations north of −60°. But in those far-off storytelling times, the celestial pole was nearer to Thuban than Polaris (see May 21). That part of the sky near the constellation Centaurus stood higher above the southern horizon than it does today. Fomalhaut and the "watery constellations" lay closer to southern seas than presently. So the products of the Greek imagination seem to us to dip further south near the Centaur than near the Water Carrier and the Whale. Still further south, the celestial sphere is figured with inventions of a modern origin. These recent additions to the stellar storybook lack the enlivening context of the Greek myths, such as those that surround the figure of Perseus.

4th: We have met Perseus before in telling the story of Andromeda. Tonight he is high overhead near the **zenith** of the sky. The constellation is not easy to pick out. The form is rambling and inchoate. If you can find Cassiopeia's big "M" and the Pleiades in Taurus, then Perseus will be lurking midway between. The two brightest stars are Mirfak (*MUR-fak*) and Algol (*AL-gall*). My drawing is adapted from a 17th-century

star atlas of Hevelius, whom we have met before as an inventor of constellations. Mirfak means "elbow," but following Hevelius I have placed the star on the hero's back. In Perseus' hand is the head of Medusa. Medusa was the Gorgon with snaky hair. She was so ugly that the direct sight of her turned the viewer to stone. Perseus slew her by chopping off her head as he looked at her reflection in his shield. He carried the head around in a bag, and used it to his own advantage for turning his enemies to stone. The star Algol is often represented as Medusa's head or eye and—as we shall see—for good reason.

DECEMBER 5

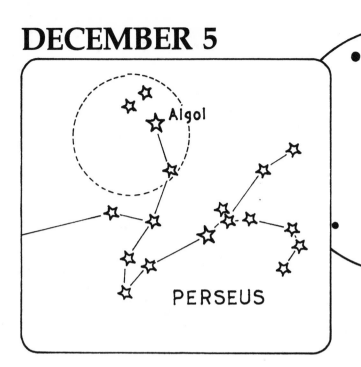

PERSEUS

DECEMBER 6

5th: Since ancient times, Algol has been considered an unlucky star. Its influence, it was believed, was invariably malevolent. The name means "the ghoul," and in myth the star is associated with the baleful head of Medusa. Algol is indeed a most unusual star. Once every 3 days it suddenly dips in brightness by more than a magnitude. It was probably this eerie "blinking" of the star that caused it to be associated with the Gorgon's evil eye. Woe to him who was caught in its stultifying stare.

6th: We know now that Algol is a prime example of an eclipsing **binary.** Algol is a double star. The more massive primary is a brilliant blue-white star several times larger than the sun. Its companion (or secondary) is an even larger yellow star, but not nearly so bright as the primary. The two stars circle each other about their common center of mass every 3 days. The plane of the motion is oriented to our line of sight in such a way that the stars eclipse one another. The sudden dip in brightness occurs when the dimmer companion passes between us and the primary, blocking the light of the brighter star. There is a less dramatic dip when the secondary is eclipsed. The two stars are only about 6 million miles apart, so close that their atmospheres intermingle. Mass may be exchanged between them. And so it is that a frightening inconstancy among stars turns out to have a rather simple explanation.

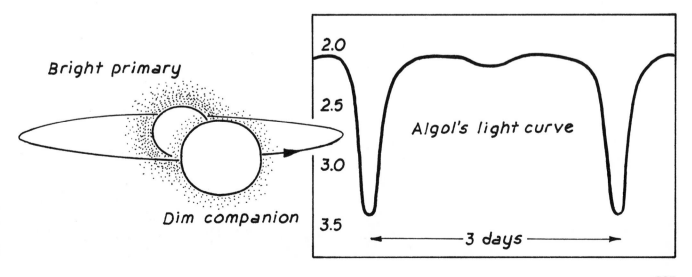

Bright primary

Dim companion

Algol's light curve

2.0

2.5

3.0

3.5

3 days

DECEMBER 7

DECEMBER 8

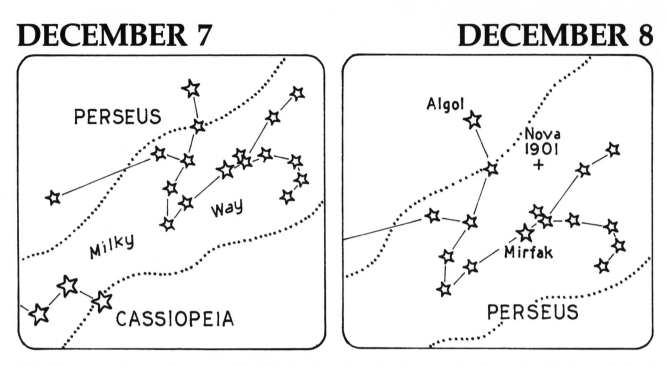

7th: The constellation Perseus lies directly astride the Milky Way. It is an area of the sky rich in stars and clusters and is well placed high overhead for scanning with binoculars on this starry night. One of the spiral arms of the Milky Way Galaxy takes its name from the constellation (see Dec. 11–12). What lies between the stars in the arms of the Milky Way? Interstellar space is filled mostly with hydrogen. "Filled" is perhaps the wrong word; the average density of the gas is about one atom per cubic centimeter, less than that of any terrestrial vacuum. This interstellar gas is best studied by radio astronomy. Hydrogen atoms emit a particular radiation with a wavelength of 21 centimeters. This wavelength is in the radio part of the spectrum. The so-called 21-centimeter radiation has been successfully used to construct a rough map of the arms of the Galaxy. The method uses the Doppler shift (see Nov. 29) and the differing rotation rates of the spiral arms.

8th: The first bright **nova** of our century appeared in Perseus in 1901. At maximum, the new star was as bright as Aldebaran and Capella. For several weeks it dropped steadily in brightness, and then went into a long period of rapid variation. Today a shell of expanding gas marks the site of the spectacular flareup. The picture below shows the sky as it might have looked near midnight on February 23, the day the nova reached maximum brightness. The view is to the northwest. For one brief night the casual observer might have thought he was looking at the summer triangle as it sets in late autumn. As we might expect, the nova was not far from the midline of the Milky Way.

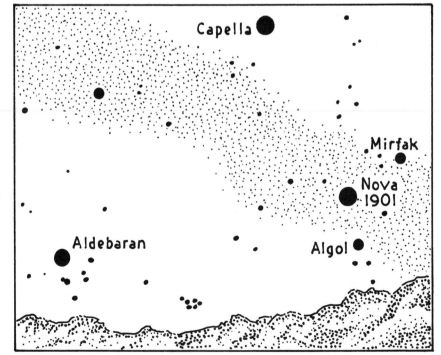

DECEMBER 9

DECEMBER 10

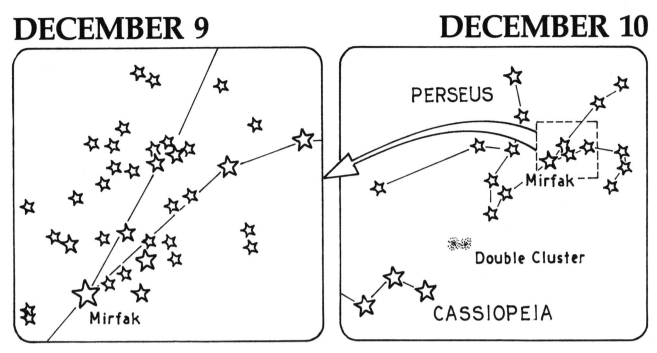

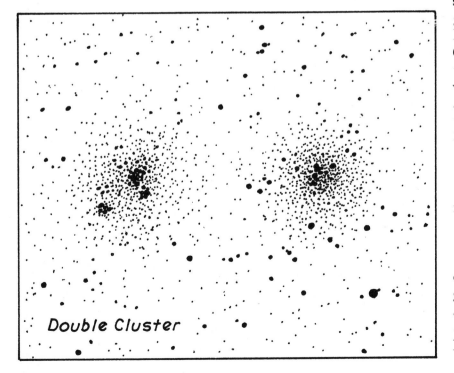

Double Cluster

9th: If you have binoculars, you will certainly want to scan the sky in Perseus. The area just to the southeast of Mirfak contains an example of what you might find. This is the so-called *Perseus III Association* of brilliant hot stars, scattered like blue diamonds on a black velvet cloth. With a good pair of binoculars you should be able to count dozens of these sparkling gems. The association is slowly dispersing. Four million years ago the cluster might have appeared as tight as the Pleiades. In the cluster there are many hot blue **main sequence** stars, which indicates that the group is very young. Yellow-white Mirfak is itself a member of the group, the brightest of the lot.

10th: Theoretically barely visible to the unaided eye, and a marvel in the eyepiece of a small telescope, is the famous *Double Cluster* in Perseus. In my opinion, this is one of the most exciting objects in the sky for the small instrument. It is one of the few such objects that did not make it into **Messier's catalogue,** and it is hard to understand why not. The clusters go by the names NGC 869 and NGC 884 (for *New General Catalogue).* The clusters are easy to find midway between the vertex of the "vee" of Perseus and the less bright end of the "M" of Cassiopeia. These two clusters are almost certainly among the youngest associations of stars in the Galaxy. Perhaps not more than a few million years have passed since they condensed from a great nebula of dust and gas. There are thousands of stars in each cluster, but we see only the great blue and white giants and a few red super-monsters. We can only guess how many smaller suns are hidden by distance.

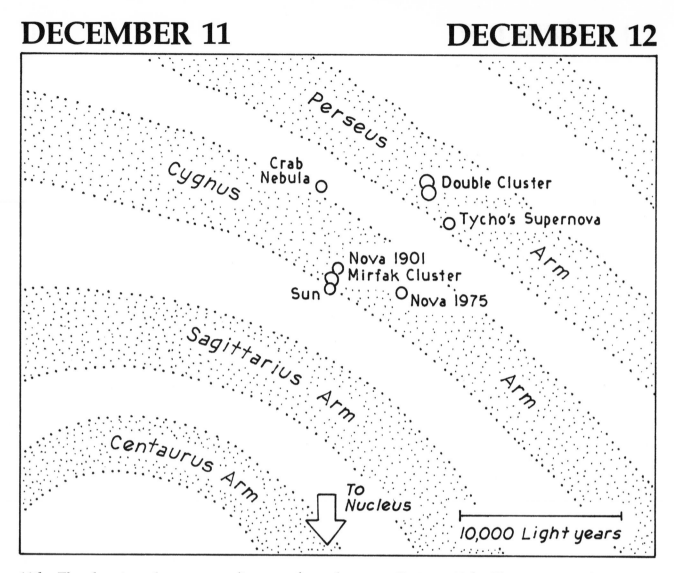

11th: The drawing above purports to show the spiral arms of the Milky Way Galaxy in the region of the sun. Actually, the detailed structure of our own galaxy is very difficult to deduce. How grand it would be to have a picture of the Milky Way Galaxy from outside; few gifts from an intergalactic star traveler would be more welcomed. In any case, the arms are not so sharply defined as in the drawing. The whole spiral whirlpool of a hundred billion stars rotates about the galactic center once every 200 million years. As you can see, the sun lies just at the inside edge of one of the spiral arms. The arm takes its name from the constellation Cygnus. The brilliant association of young stars near Mirfak lies in the Cygnus Arm. Most of the dust and gas of the galaxy is confined to the arms, and it is here that we would expect to find associations of hot young stars. The arms are prolific incubators of new stars. The arms are also defined by the occasional **novae** or **supernovae** that announce the deaths of stars. Several of the novae that we have studied on previous starry nights were flare-ups of old stars that today lie hidden in the thick star streams of the Cygnus Arm.

12th: The next spiral arm out beyond our own takes its name from the constellation we have been observing on recent starry nights. The beautiful Double Cluster in Perseus offers a rare naked-eye glimpse of this distant realm of stars. Much of our picture of the spiral arms of the Galaxy is derived from the work of radio astronomers (see Dec. 7). But a careful study of the positions of clusters and star associations, and of young O-type giants and red supergiants, also reveals the rough outlines of the spiral. The star death witnessed by Tycho Brahe in 1572 (see Nov. 7) occurred in the Perseus Arm.

DECEMBER 13

DECEMBER 14

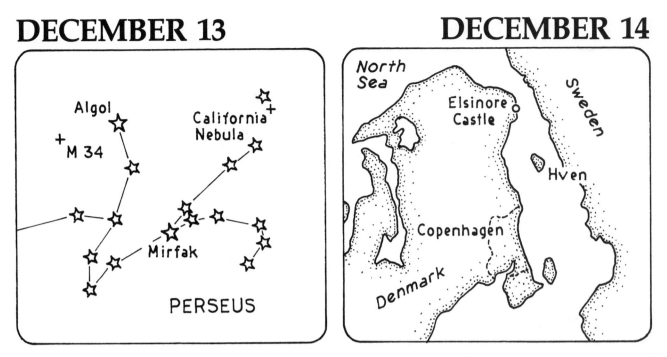

PERSEUS

13th: As we might expect for a constellation that lies astride the Milky Way, there is much to look for in Perseus. Let us take a last glance at two more interesting objects. The first is the *California Nebula*, which takes its name from a shape that is supposed to look like the western state. It takes, I think, some stretch of the imagination to recognize the similarity. But for the fact that the nebula has a prominent "San Francisco Bay," I doubt if it would have received its name. But never mind, California is the only one of the United States with its name immortalized among the stars. It is also the only state for which an atomic element has been named, and can therefore take some pride of place at both ends of the scale of nature. The faint light of the California Nebula can only be recorded on long-exposure photographs. M 34, on the other hand, is an object that might reveal itself to binoculars. It is a loose cluster of stars, not so spectacular as the Double Cluster or the Mirfak association, but well worth looking for.

14th: December is a month of birthdays for famous astronomers. The first, which we celebrate today, is that of the great Danish observer Tycho Brahe (see Nov. 7). For 25 years, Tycho studied the starry night from his private observatory on the Isle of Hven in the strait between Denmark and Sweden, just south of Hamlet's Castle at Elsinore. Tycho was the most careful of naked-eye observers (his work came just before the invention of the telescope). The most prominent crater on the moon has been named for him.

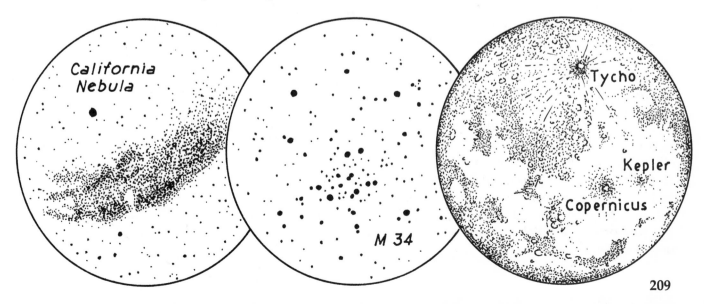

California Nebula

M 34

Tycho

Kepler

Copernicus

Mirfak

Algol

Northwest

Algol

Mirfak

North

15th: Tonight at 10 o'clock, the demon star Algol is almost exactly overhead for an observer in the central United States. If the influence of "the ghoul" is indeed malevolent, no time should be so unlucky as when the star is at the **zenith.** At the same moment, observers at other locations on the planet have a very different view of Algol, and perhaps a less fearsome glance from the Medusa's eye. The drawing of the earth shows the nighttime side of the planet at the moment this evening when Algol is overhead for our U.S. observer. In Asia it is daytime. Near Rome, dawn is not far away. An Italian stargazer sees Algol near the northwestern horizon, as in the drawing above. As time passes, the star—the entire constellation —will slide down toward the north and set behind the sea cliffs. Imagine a plane tangent to the planet at the observer's feet. This is the plane of the observer's horizon, and defines the half of the celestial sphere that is visible.

16th: At the same moment when Algol is overhead in the United States, an observer on the western coast of South America sees the star low on the northern horizon, and slipping down toward the northwest. The **zenith** star for our South American observer is Rigel, and the season is summer.

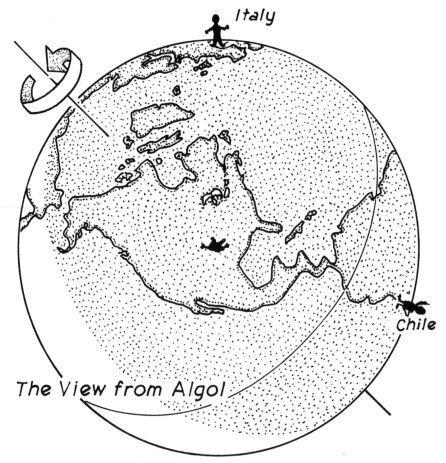

Italy

The View from Algol

Chile

DECEMBER 17

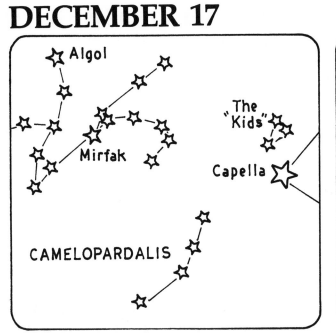

DECEMBER 18

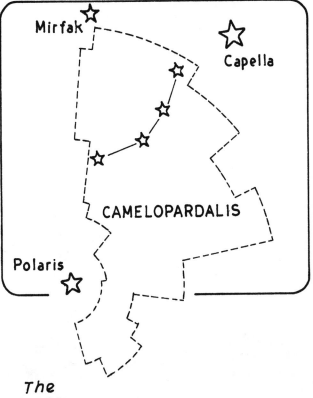

The Giraffe

17th: Camelopardalis *(ka-MEL-o-PAR-da-lis)* the Giraffe is certainly the most unfamiliar of the large constellations in the northern sky. Perhaps its greatest claim to our attention is precisely the fact that its borders include so spacious and prominent an area of the sky, and yet few star books mention it. Camelopardalis (or sometimes simply Camelopardus) contains no bright stars, in fact none that can be seen on a typical city or suburban night. Most of the black gulf that falls between Capella and Polaris belongs to the constellation. The Giraffe faces the Great Bear in the sky, a rather precarious position for so gentle a creature. Almost half the 88 official constellations are nonhuman animals, and if gathered together would make a fine small zoo. The zoo would include several creatures not found in any zoo you have visited, such as the phoenix and the unicorn. Would the centaur be an exhibit or a spectator?

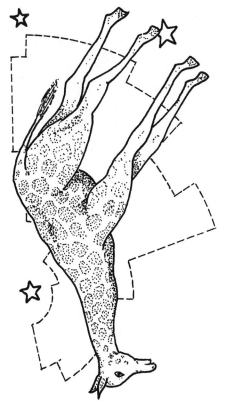

18th: The Giraffe is a modern constellation, another of the compulsive "mopping-up operations" that accompanied the birth of modern science in the 16th and 17th centuries. This was also a time when Europeans were becoming familiar with the strange creatures of Africa (recall, for example, Durer's wonderful woodcut of a rhinoceros). Only the elongated outline of Camelopardalis suggests a giraffe. The official name of the constellation is given in the Latin form, as for most constellations. Even those constellation names with Greek origins have been Latinized. Only Dorado, a constellation near the south celestial pole (and which includes the Large Magellanic Cloud) is an exception to this rule. The name is Spanish and means "goldfish."

DECEMBER 19

DECEMBER 20

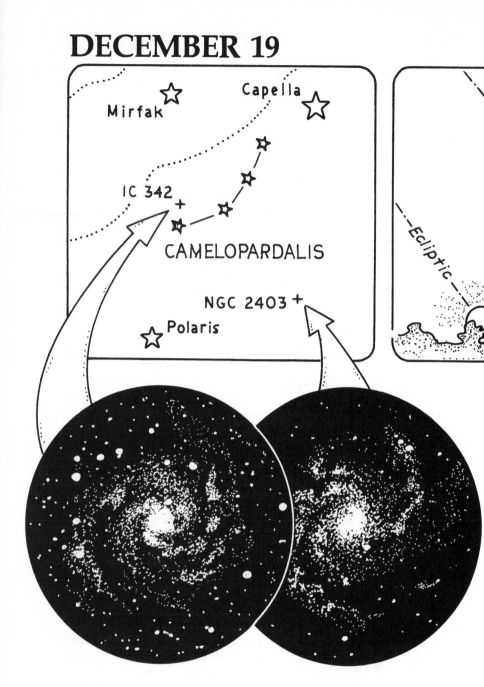

19th: Two of our nearest galactic neighbors are in Camelopardalis. IC 342 (IC for the *Index Catalogue* of 1895) is almost hidden in the outlying fringes of the Milky Way (the stream of light, not the Galaxy). It is a face-on spiral, another heavenly pinwheel. The obscuring matter of the Milky Way makes it difficult to get an accurate distance to IC 342, but it is probably a member of the **Local Group.** NGC 2403 is a large spiral with poorly defined arms. It is one of the nearest galaxies beyond the Local Group. At 8 million light years, it is four times further away than the Great Galaxy in Andromeda. Though more distant than IC 342, the light of NGC 2403 is concentrated into a smaller area of the sky and the galaxy might possibly be glimpsed with binoculars. This is one of the most distant objects you can see with such an instrument.

20th: We are approaching the **winter solstice.** The sun reaches its southernmost excursion on the **ecliptic** sometime near December 21 or 22. The precise moment of the solstice varies from year to year. If you have been watching the sun go down in the west night by night, you will have noticed the place of sunset has been creeping southward along the horizon. On the next few nights the sun will set at its most southerly point. Compare, for example, the drawing above to the one on December 1. How far south of west the sun sets depends on your latitude. If you live far enough north, above the arctic circle, the sun will not rise at all! The solstice is the day when the northern hemisphere of the earth receives its smallest share of the sun's radiation. It is not, of course, the coldest day of the year. The earth still retains some of summer's warmth and the coldest starry nights will be reserved for January.

DECEMBER 21

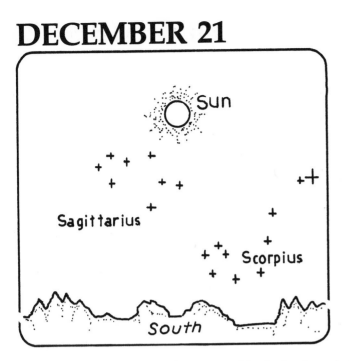

DECEMBER 22

21st: At noon today the sun stands as close to the southern horizon as it ever gets. The drawing above shows the view to the south at noon and indicates the constellations that are presently in that part of the sky, hidden in the sun's light. If the sun were briefly eclipsed and the stars came out we would see that the sun stood near the knob of the teapot. The **winter solstice** is in Sagittarius.

22nd: At the winter solstice the sun stands directly above the **Tropic of Capricorn,** 23½° south of the equator. Thousands of years ago the sun stood in Capricorn, not Sagittarius, on the day of the solstice. The sun's light now strikes the northern hemisphere of the earth at a long, slanting angle, and spreads its warming effect over a larger area of the surface. This dilution of the sun's radiation, not the slightly varying

distance of the earth from the sun, causes the change in seasons. If the earth's axis were not tipped to the plane of its orbit, our lives would be very different. We would have climates (cold poles, warm equatorial regions), but no seasons. Trees would not shed their leaves, birds would not migrate, bears would not hibernate. Starry nights would feel like spring and autumn all the year round.

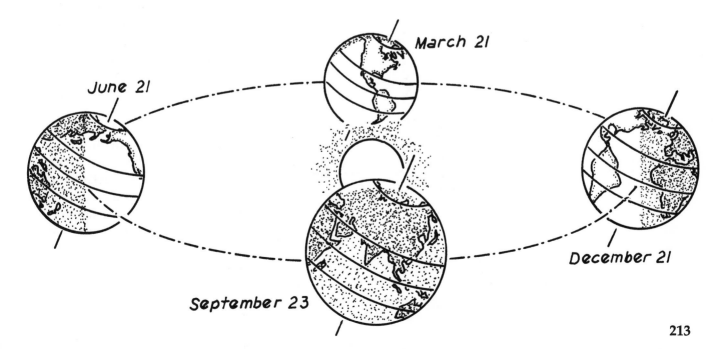

DECEMBER 23

DECEMBER 24

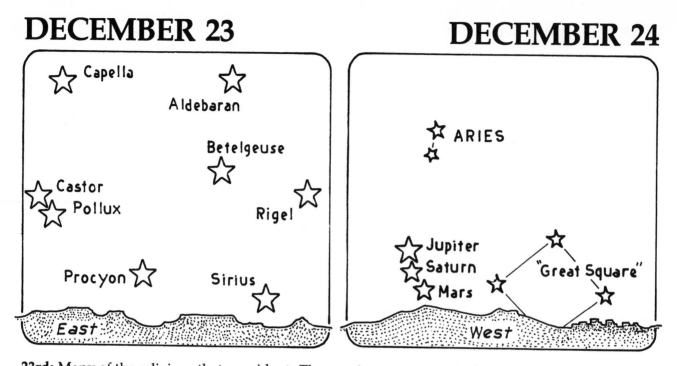

23rd: Many of the religions that grew up in the northern hemisphere celebrate "feasts of light" in December. The Jewish *Chanukah* celebration lasts for eight days of the month, with a new candle lit in a special menorah on each day of the festival. Candles (or, these days, electric lights) also play an important role in the Christian celebration of Christmas. The near-coincidence of these feasts with the **winter solstice** is no accident. The sun is now at its weakest for those of us who live north of the equator. Quite aside from the religious significance, the lighting of candles might have once been a symbolic attempt to help rekindle the sun's strength. A glance at the eastern horizon this starry night suggests another natural complement for the "feasts of light." The sky is suddenly ablaze with "night candles."

24th: Christmas Eve! The one night when I would prefer a snowfall to stars. But stars have played an important symbolic role in celebrations of Christmas. Perhaps the most famous star of all is the Christmas Star that the Magi followed to Bethlehem. Was there a real Christmas Star, and if so, what was it? There have been many theories. Guy Ottewell has a nice discussion of the possibilities in the book I recommend in "Sources and Resources." Jesus was probably born sometime between 7 and 4 B.C. During this period a splendid candidate for the Star of Bethlehem was the conjunction of Jupiter, Saturn, and Mars in February of 6 B.C. An equally dramatic grouping of the three planets occurs only once every 800 years. The prospect of the planets blazing and dancing together on the western horizon might have seemed an omen of a great event. The drawing above shows the apparition that the three kings might have followed to the stable in Judea.

DECEMBER 25

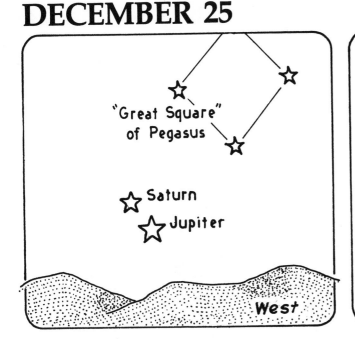

"Great Square" of Pegasus

☆ Saturn

☆ Jupiter

West

25th: Jupiter and Saturn overtake each other in the sky once every 20 years. They were again in conjunction near the border of Pisces and Aquarius during the winter of 1642. The drawing above shows the sky as it might have looked from the village of Woolsthrope in Lincolnshire, England, on the evening of Isaac Newton's birth on Christmas Day 1642, almost exactly one year after the death of Galileo. It was Newton, perhaps more than any other man, who unraveled the mysteries of the starry night.

DECEMBER 26

26th: The apple and the moon! Newton's great work—*The Mathematical Principles of Natural Philosophy*, better known by the abbreviated Latin name *Principia*—was published in 1687. In this book Newton laid out a complete mathematical theory of gravitation. It is said that the germ of the theory occurred to Newton when an apple fell from a tree as he sat in the garden of his house at Woolsthrope. *Could the force that pulled the apple to earth be the same as the force that held the moon in orbit around the earth?* Beginning with this insight, Newton went on to create his theory of gravity, and to calculate the subtle motions of apples, moons, planets, tides, and stars. "Tension!" exclaims naturalist May Theilgaard Watts, "It can hold a poem together; make a violin string sing; shape the tides; enable suspenders to hold up pants against the pull of gravity; and is the essence of the resilient, bouyant landscape where forces ever pull against each other." The tension between nuclear fusion and gravity gives life to a star. The tension between rotation and gravity gives shape to a galaxy. Gravity is one term of the universal tension that makes the universe sing.

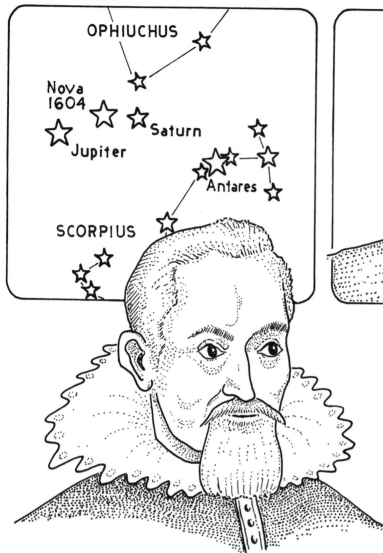

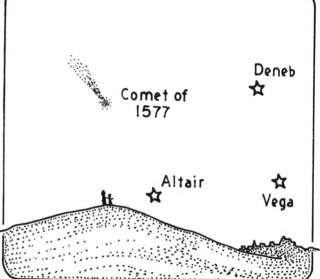

27th: The third of our December birthdays is that of Johann Kepler, the brilliant protégé of Tycho Brahe. Tycho was the older of the two men and a masterful observer. He amassed a wealth of precise astronomical data, including careful records of the motions of the planets. It was Kepler who provided a theoretical understanding of Tycho's data. Kepler was obsessed all his life with a search for laws of mathematical harmony in the universe. From Tycho's records on the planetary positions, he was able to deduce what we now call *Kepler's laws of planetary motion*. It was Kepler who first recognized that planets move in elliptical orbits. And he found simple mathematical laws describing the speeds of the planets in their orbits. Kepler's laws would later be used by Newton as an inspiration and test for his *theory of universal gravitation*. Tycho Brahe and Johann Kepler can be taken to symbolize the two foundation stones of modern astronomy: exact observation and mathematical analysis.

28th: When Kepler was 5 years old, his mother took him outside to view a bright comet that had appeared in the sky. This was the Great Comet of 1577 which Tycho Brahe studied at Hven. Tycho's contemporaries believed that the universe beyond the moon was made of immutable crystalline spheres, centered on the earth. The comet, they thought, must be some sort of distillation of the earth's atmosphere, a divine omen or portent. From his observations, Tycho was able to demonstrate that the comet was more distant than the moon, and that therefore the heavens were subject to change after all. The comet was the first dramatic link between the young Kepler and the older Dane. It may also have been the spark that ignited in the 5-year-old boy's mind a lifelong passion for the sky. It was Kepler's work, founded on Tycho's data and expanded by Newton, that removed comets from the realm of superstition and placed them in the context of natural law.

216

DECEMBER 29

DECEMBER 30

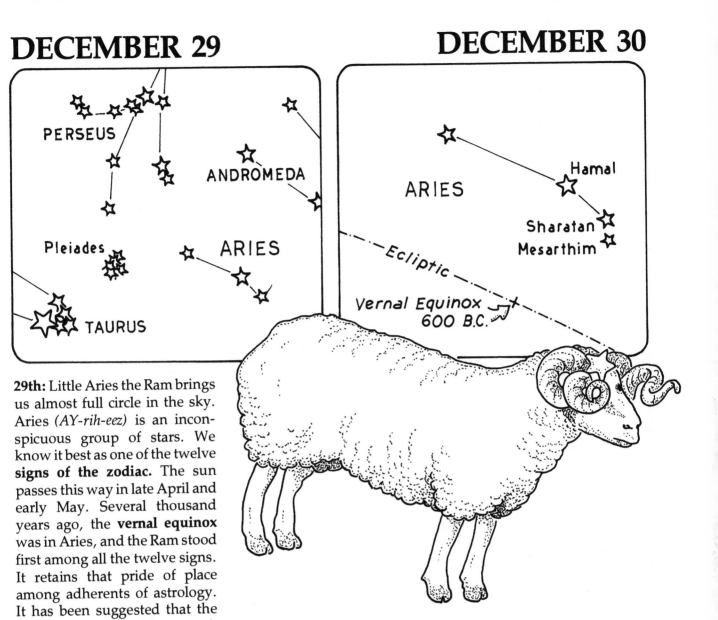

PERSEUS

ANDROMEDA

Pleiades

ARIES

TAURUS

ARIES

Hamal

Sharatan

Mesarthim

Ecliptic

Vernal Equinox 600 B.C.

29th: Little Aries the Ram brings us almost full circle in the sky. Aries *(AY-rih-eez)* is an inconspicuous group of stars. We know it best as one of the twelve **signs of the zodiac.** The sun passes this way in late April and early May. Several thousand years ago, the **vernal equinox** was in Aries, and the Ram stood first among all the twelve signs. It retains that pride of place among adherents of astrology. It has been suggested that the Babylonians created the constellation when the equinox moved into this part of the sky, taking stars from Taurus and Pegasus to do so. This supposedly accounts for the fact that bull and winged horse are traditionally shown without the back parts of their bodies. As we have seen, the spring equinox has now moved on into Pisces, but it is still referred to as the "first point of Aries," or simply "Aries." This is just one anachronism which the **wobble** of the earth's axis has introduced into astronomy.

30th: Hamal is the most prominent of the very "unprominent" stars of Aries. The name means "ram." The star is an orange giant and lies 75 light years from earth. Sharatan, the "sign," has a proper name of its own mainly because it once stood close above the spring equinox. Mesarthim, the "fat ram," was also once the "signpost" that marked the position of the sun as it crossed the sky's equator. According to the myths of the Greeks, Aries was the ram of the famous golden fleece. If so, we must assume that when the ram was sacrificed and the fleece removed by Phrixus, only the bones were placed in the heavens. The constellation has too little "shine" to represent a fleece of gold. Just south of Aries is the tail of Cetus, thrashing in the waters of the river Eridanus. The river is not an easy constellation for northern observers, flowing from a spring near Orion's foot down over the southern horizon to the first-magnitude star Achernar ("end of the river").

DECEMBER 31

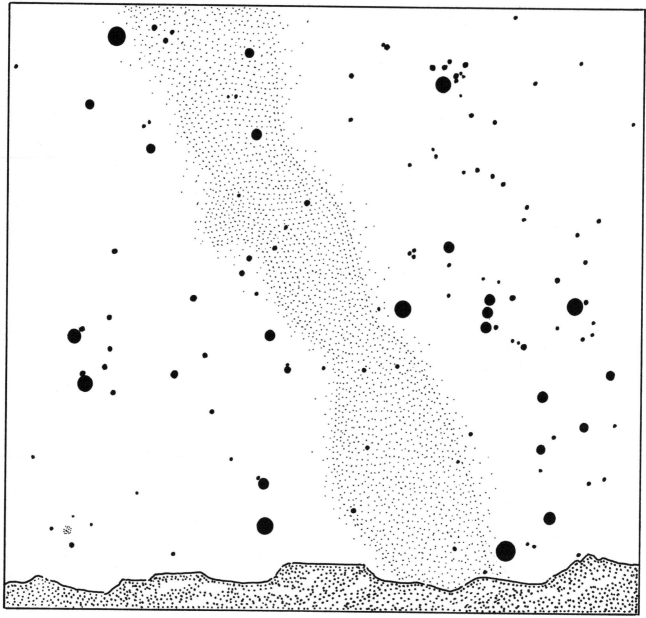

31st: Crossing the meandering Eridanus we return to the constellations with which we began the year. The drawing shows the sky in the east this evening, ablaze with stars. How many of the stars and constellations can you name? There are many reasons to learn the stars. In their names and in their fanciful arrangements are a history of the imagination of humankind. To this add the joy that comes from following in the sky the changing seasons. Arcturus rising is as sure a harbinger of spring as the robin. Orion returning with winter's chill is an old and faithful friend. And then there is the sheer pleasure of recognition, of saying, "This is the Eagle, and that is the Swan." The sky is half our visual field; to ignore it is to cut oneself off from half of the world's beauty. The Little Prince in Saint-Exupery's story lived on a planet so small he could watch the sun set a dozen times in a single evening simply by moving his chair. We are not so lucky. We must wait and let the sky come to us. And so we have waited and watched through a year of starry nights. At the stroke of midnight, as the New Year begins, the stars in the drawing will be high overhead. Watch for Sirius to pass exactly south of where you stand. That will be the moment to begin your New Year's revels.

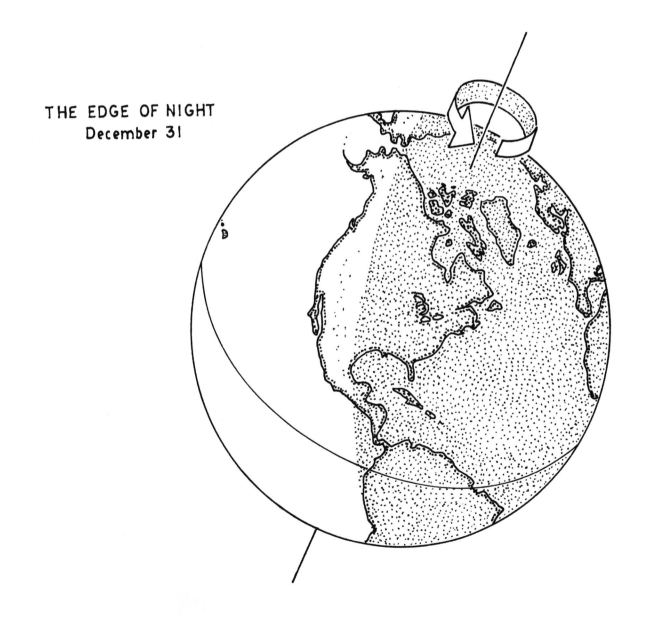

THE EDGE OF NIGHT
December 31

Sources and Resources

A short list of atlases and other books you may like to know about.

A. Becvar, *Atlas of the Heavens* (Prague: Publishing House of the Czechoslovak Academy of Sciences, 1962).
The most beautiful buy you will ever make for your money. This colorful atlas is a delight to mind and eye. Useful for yourself and a perfect gift for anyone who loves the sky. Can be obtained from Sky Publishing Corp., 49 Bay State Rd., Cambridge, Mass. 02238. You should certainly write for Sky's catalogue of publications. They distribute many books and atlases which I have found

useful and which you may want to own.

R. Burnham, Jr., *Burnham's Celestial Handbook: An Observer's Guide to the Universe beyond the Solar System*, 3 volumes (New York, N.Y.: Dover, 1978).
Burnham has collected in one place an astonishing amount of information. I found myself often referring to Burnham's book as I prepared this one. Material is organized by constellation and covers the entire celestial sphere. If my book whetted your appetite, you can feast on Burnham's.

T. Dickinson and S. Brown, *The Edmund Sky Guide* (Barrington, N.J.: Edmund Scientific Co., 1977).

221

A companion to the *Edmund Mag 5 Star Atlas*. If you move on to observing the sky with binoculars or telescope, these two little works will be the best first additions to your library.

R.A. GALLANT, *The Constellations: How They Came to Be* (New York: Four Winds Press, 1979).
A very readable retelling of the myths and lore that make the constellations come alive.

A. P. NORTON, *Norton's Sky Atlas and Reference Handbook* (Cambridge, Mass.: Sky Publishing Corp., 1973).

A compact and useful reference work for the serious star-gazer. This classic (first published in 1910) remains the best of its kind.

G. OTTEWELL, *The Astronomical Companion* (published by author).
This is a complement to Ottewell's annual *Astronomical Calendar* of celestial events. Both the *Companion* and the *Calendar* are unique works, imaginatively illustrated, which you will certainly want to own. The annual *Calendar* will greatly enhance your enjoyment of the starry night. The

Companion is an inexpensive grand tour of the universe. The books can be obtained c/o Dept. of Physics, Furman University, Greenville, SC 29613. Highly recommended.

H. A. REY, *The Stars: A New Way to See Them* (Boston: Houghton Mifflin Co., 1976).
Rey's delightful books have been favorites of stargazers for a generation. Young people especially will appreciate his lucid prose and charming drawings.

Glossary

A list of terms used on more than one occasion, sometimes without definition.

altitude: Angular distance of a celestial object above (or below) the horizon, measured perpendicular to the horizon. *Oct. 14*

apparent magnitude: A measure of the brightness of a celestial object as seen from earth; brighter objects have smaller numerical magnitudes. *Jan. 9, Oct. 26*

autumnal equinox: The point on the celestial equator where the sun crosses the equator moving south; also the time when the sun is there. *Oct. 19*

azimuth: The angular distance from north measured east-ward along the horizon to a point vertically below a celestial object. *Oct. 13*

binary star: A double-star system—two stars revolving about a common center of mass. *Mar. 17*

black hole: A collapsed massive star, so dense that not even light can escape the pull of its gravity. *Jan. 31*

celestial equator: Points on the celestial sphere directly above the equator of the earth. *Jan. 3, Oct. 9*

celestial poles: The two points on the celestial sphere directly above the poles of the earth; the intersection of the earth's axis with the celestial sphere. *May 18, Oct. 9*

celestial sphere: The imaginary sphere of the sky on which all celestial objects apparently reside; believed to literally exist in ancient times. *Jan. 4, Oct. 25*

cepheid variable: A type of bright giant star that shows regular characteristic variations in brightness; used as a distance indicator. *Sept. 16–19*

circumpolar stars: Those stars closer to the celestial pole than the horizon, and therefore which never set. *May 2, Nov. 4*

declination: Angular distance of a celestial object north or south of the celestial equator; analogous to the latitude of a terrestrial location. *Oct. 9*

ecliptic: The sun's apparent annual path across the celestial sphere. *Jan. 19*

ecliptic plane: The plane of the earth's orbit about the sun; approximately the plane of the entire solar system. *Jan. 20, May 22*

equator: See *celestial equator*.

equatorial coordinates: The coordinates (right ascension and declination) that describe the position of an object on the celestial sphere with respect to the celestial equator and the vernal equinox. *Oct. 9*

galactic cluster: See *open cluster*.

galactic equator: The line in the sky that corresponds to the central plane of the Milky Way Galaxy. *July 31*

galactic poles: The points on the celestial sphere 90° from the galactic equator; the intersection on the celestial sphere of a line through the observer parallel to the axis of the Milky Way Galaxy. *Apr. 29*

galaxy: A large system of millions to hundreds of billions of stars, sometimes containing large amounts of dust and gas.

Galaxy: The galaxy to which the sun belongs and which we see as the Milky Way. *July 21*

globular cluster: A spherical distribution of tens of thousands of stars gravitationally bound to a galaxy. *Mar. 25*

H-R diagram: The Hertzsprung-Russell diagram; a graph showing the absolute brightness (luminosity) of stars as a function of their temperature (or color or spectral type). *Mar. 15*

horizon coordinates: The coordinates (azimuth and altitude) of a celestial object with respect to the observer's horizon. *Oct. 14*

light year: The distance light travels in a year—about 6 trillion miles. *Jan. 21*

Local Group: The cluster of galaxies to which the Milky Way Galaxy belongs. *Nov. 18*

luminosity: The rate at which a star radiates energy; the intrinsic brightness of a star. *Mar. 15*

magnitude (apparent): See *apparent magnitude*.

main sequence: A diagonal band on the H-R diagram which describes stars during the normal course of their stable hydrogen-burning lifetimes. *Mar. 15*

Messier's catalogue: A catalogue of fixed nebulous nonstellar objects compiled by Charles Messier in 1787. *Mar. 23*

Messier object: An object listed in Messier's catalogue.

meteor: A track of light in the sky resulting from the vaporization of solid matter entering the earth's atmosphere; commonly called "shooting stars" or "falling stars." *Apr. 11*

Milky Way: The band of faint light circling the celestial sphere that has its source in the myriad stars of the Milky Way Galaxy. *Feb. 19, July 21*

neutron star: A collapsed extremely dense star consisting almost entirely of neutrons; the final state of a star about twice as massive as the sun. *Jan. 28*

nova: A star that suddenly flares in brightness by a factor of hundreds of thousands. *July 14*

occultation: The eclipse of one celestial object by another, as when the moon or a planet passes in front of a star. *Jan. 20*

open cluster (galactic cluster): A loose cluster of stars of common origin, found in the spiral arms of the Galaxy. *Jan. 25, Mar. 13*

parallax: The apparent change in the position of an object when viewed from two different locations; used to measure the distance to nearby stars. *June 4*

planetary nebula: A thick shell or ring of gas moving out from a star; probably the outer layers of the star blown off during the star's dying contractions. *Feb. 28*

pole: See *celestial pole.*

precession: The slow wobble of the earth's axis which causes the position of the celestial poles and equator to change relative to the stars. *Mar. 8, May 22*

proper motion: Motion of a star across the sky relative to the positions of the distant stars and galaxies. *Feb. 17*

pulsar: A celestial object that emits radio energy (sometimes light) in short regular bursts; probably a rapidly rotating neutron star. *Jan. 28*

quasar (quasi-stellar radio source): An intense starlike source of radio energy and light, characterized by a large red shift and therefore probably very distant; the nature of these objects is poorly understood. *May 30–31*

radiant (of meteor shower): The point in the sky from which meteors in a shower seem to radiate. *Apr. 11*

red dwarf: A small cool star at the bottom of the main sequence. *Apr. 15*

red giant: A cool red star, very bright because of its large size; a late stage in the lifetime of a typical star. *Jan. 15–16, Mar. 15*

right ascension: Angular distance of a celestial object measured eastward along the equator from the vernal equinox; analogous to the longitude of a terrestrial location. *Oct. 10*

sign of the zodiac: An astrological term; one of twelve equal segments of the ecliptic, corresponding to the 12 constellations of the zodiac. *June 25*

spectral type: A classification of stars based on an analysis of their spectra; in general, each spectral type corresponds to a range of temperature. *Feb. 9*

spring equinox: See *vernal equinox.*

summer solstice: The point on the celestial sphere where the sun reaches its northernmost declination; also the time when the sun is there. *Feb. 25, June 21*

supernova: An unusually violent explosion of a star, which results in an increase in brightness of hundreds of millions of times. *Jan. 26, July 14*

Tropic of Cancer: Parallel of latitude on earth, 23½° north of the equator, corresponding to the sun's northernmost declination. *Mar. 7*

Tropic of Capricorn: Parallel of latitude on earth, 23½°

south of the equator, corresponding to the sun's southernmost declination. *Mar. 7, Sept. 24, Dec. 22*

vernal equinox: The point on the celestial equator where the sun crosses the equator moving north; also the time when the sun is there. *Apr. 2*

white dwarf: A small dense hot white star; the final stage in the evolution of stars with masses similar to the sun's or smaller. *Jan. 27, Feb. 14, Mar. 15*

winter solstice: The point on the celestial sphere where the sun reaches its southernmost declination; also the time when the sun is there. *Feb. 25, June 21*

"wobble" (of earth's axis): See *precession.*

zenith: The point in the sky directly over an observer's head. *Jan. 1*

zodiac: The twelve constellations that the sun passes through during its annual journey around the celestial sphere. *Jan. 20, June 25*